THE NEW BEST OF FINE WOODWORKING

Workshop Machines

THE NEW BEST OF FINE WOODWORKING

Workshop Machines

The Editors of
Fine Woodworking

The Taunton Press, Inc., 63 South Main Street, PO Box 5506, Newtown, CT 06470-5506
e-mail: tp@taunton.com

Jacket/cover design: Susan Fazekas
Interior design: Susan Fazekas
Layout: Susan Lampe-Wilson
Front cover photographer: Tom Begnal, courtesy *Fine Woodworking*
Back cover photographers: (left) Robert Vaughan, courtesy *Fine Woodworking*;
(right, top and bottom) Alec Waters, courtesy *Fine Woodworking*

Library of Congress Cataloging-in-Publication Data

Workshop machines : the new best of Fine woodworking / the editors of
Fine woodworking.
p. cm. -- (Best of fine woodworking)
ISBN 1-56158-765-6
1. Woodwork. 2. Furniture making. I. Taunton's fine woodworking. II. Series.
TT180.W695 2005
684'.083--dc22
2004018513

Printed in the United States of America
10 9 8 7 6 5 4 3 2 1

Working wood is inherently dangerous. Using hand or power tools improperly or ignoring safety practices can lead to permanent injury or even death. Don't try to perform operations you learn about here (or elsewhere) unless you're certain they are safe for you. If something about an operation doesn't feel right, don't do it. Look for another way. We want you to enjoy the craft, so please keep safety foremost in your mind whenever you're in the shop.

Acknowledgments

Special thanks to the authors, editors, art directors, copy editors, and other staff members of *Fine Woodworking* who contributed to the development of the articles in this book.

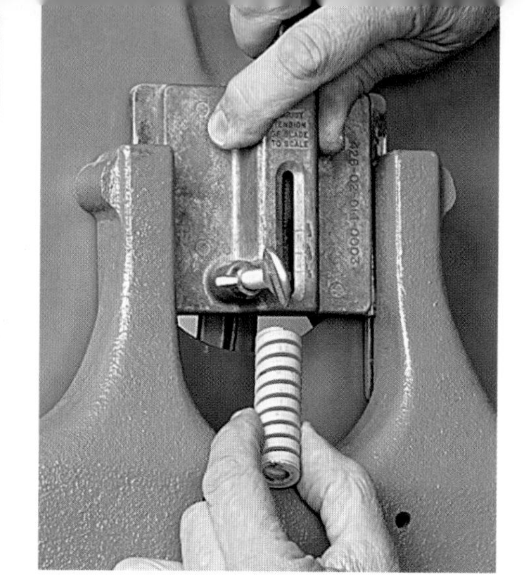

Contents

Introduction

My first major machine-tool purchase was a garage-sale tablesaw. The tool was all cast iron and sheet metal. There was very little rust to be found, suggesting a well-cared-for history, and it appeared to include all of the necessary parts. When I hit the power switch, the motor came to life with a powerful, satisfying hum. But try as I might, I was unable to rip a board with parallel edges or crosscut stock squarely.

The owner's manual was not much help. It told me how to assemble the various parts and suggested I should not operate the tool while standing in a puddle of water. As I went over the machine, I realized that my ignorance was not limited to technique: This saw was in need of a tune-up.

I wish I had this book then. Over the years, *Fine Woodworking* magazine has published a number of articles on how to use machines safely and efficiently and what to do when they get out of tune. This book is a collection of those articles, including such topics as how to straighten warped lumber on a jointer, choose

the best bandsaw blade for a particular job, or set the blade of a tablesaw parallel to the miter-gauge slot.

Whether you are just starting out in woodworking and need instruction on the basics of how to use machines to get flat, square stock, or need more advanced tips on useful jigs for your machines, this book will take your woodworking to a higher level.

—Anatole Burkin,
Editor of *Fine Woodworking*

THE THICKNESS PLANER IS an essential tool for milling wood. It will quickly and accurately give you a second flat face, parallel to the jointed face.

Flat, Straight, and Square

BY PETER KORN

If wood were a stable, homogenous, man-made material like metal or plastic, how much simpler the woodworker's task would be! Instead, we pay a price for our love of natural beauty. We work with a biological medium that reacts continuously to the environment, changing dimension and shape as it breathes moisture in and out.

If you've ever attempted to dovetail cupped boards or to build a frame-and-panel door from bowed lumber, you know how frustrating it is to work with poorly prepared stock. Fine craftsmanship occurs one step at a time, and the first step is preparing (milling) stock foursquare—straight, true, and accurately dimensioned.

Boards are almost never flat enough to use directly from the lumberyard. Even S2S (surfaced two sides) lumber is milled only with thickness planers, which create boards of uniform thickness but do little to iron out cup, bow, or twist. In any case, wood's propensity for continuous movement dictates milling only when you're ready to use it. Stored wood will often warp between milling and joinery.

I begin every project with a cutting list specifying the exact measurements of all the parts, including allowances for tenon length. If there are curved parts, I usually mill foursquare blanks and then bandsaw the curves later. Before cutting, I lay out the location of each part on the rough lumber with chalk or a crayon, trying to minimize waste and make the most attractive use of grain. Laying out the parts right on the milled stock also prevents embarrassing mistakes.

When the cutting list calls for several parts of the same dimension, you must weigh time against conservation of material. It takes less time to mill long pieces to thickness and then cut shorter parts from those long boards than it does to crosscut smaller pieces and then mill them. Longer boards tend to be more twisted and/or bowed along the length of their faces. If you mill a long board whole, however, you'll lose more material in making it flat than if you'd cut it into shorter rough blanks. The right approach is always a judgment call, depending on how straight your rough lumber is and how much thickness you have to spare between rough lumber and the dimension of your finished stock.

When I cut rough lumber into blanks for milling, I leave the blanks at least ¼ in. wider and 1 in. longer than the final pieces I'll need. I crosscut to rough length with a radial-arm saw, circular saw, or handsaw depending on what's handy and where I am. For the initial rip to rough width, I prefer the bandsaw (see the photo at right) because it's quieter, less dust-producing, and safer than the tablesaw. Where the cups, bows, and twists endemic to rough lumber increase the likelihood of tablesaw kickback, a bandsaw purrs right on through. I bandsaw by eye to the lines I have marked.

If you prefer to rip rough lumber to width on the tablesaw, make sure the edge against the rip fence is straight. If it's not, run it over the jointer first.

RIPPING ROUGH STOCK on the bandsaw is far safer than on the tablesaw because there's no danger of kickback if a warped or twisted board shifts as it's going through the blade. The author rips to about ¼ in. more than finish width to allow for jointing an edge and then ripping parallel to that trued edge.

Six Steps to Foursquare Stock

These are the steps I use to prepare stock from rough lumber:

Step 1: Flatten the first face of the board.
Step 2: Make the second face of the board parallel to the first face at the desired thickness.
Step 3: Square an edge. (Steps 2 and 3 are often reversed.)
Step 4: Rip the second edge square and to the desired width.
Step 5: Cut one end square.
Step 6: Cut the other end square and to length. (Usually, I leave stock an inch or so long until the joinery is cut.)

Before power tools existed, the entire milling process was done exclusively with hand tools, but milling is one job that machines do much more efficiently than

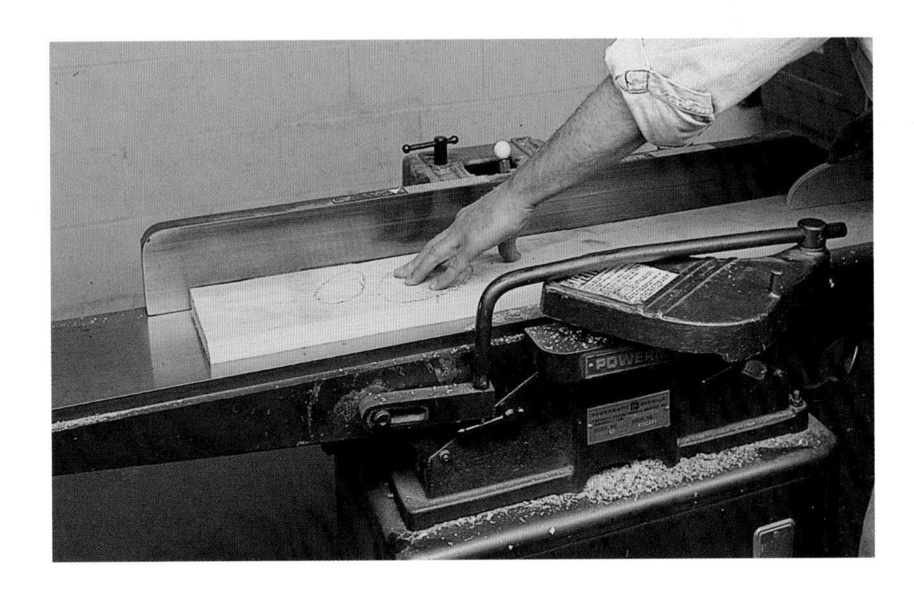

THE FIRST STEP IN STOCK PREPARATION is to create one flat face, which becomes the reference face. So long as your jointer's capacity is greater than the width of the board you need to flatten, it's a quick operation. Whenever you joint, maintain pressure on the outfeed table just past the cutterhead. Always use a push stick.

Flattening the Face of a Board with a Handplane

Like most other woodworking skills, handplaning wide boards is daunting only until you give it a serious try. All you need are one or two well-tuned planes, some elbow grease, and a couple of days of practice.

The two planes I use are a scrub plane and a bench plane. The bench plane alone would be sufficient, but the scrub plane saves time and effort by removing wood quickly from seriously cupped boards. Scrub planes are made for rough work, and there is no trick to tuning them beyond sharpening the curved blade. For flattening boards, I prefer a 14-in. jack plane, but I know other woodworkers who like to work with a longer, heavier plane, such as a 22-in. jointer plane.

To set up for planing, I clamp the work on a flat benchtop, making sure that nothing extends above the board's surface. I usually place the convex side up, so the board sits securely on the bench. Also, I find it easier to take down a center hump than to attack raised edges.

Using winding sticks and a long straight edge, I locate and mark the board's high spots and determine the degree of twist. Winding sticks are simply sticks of uniform width placed across each end of the work, parallel to each other. Sighting along their tops reveals the exact degree of twist in a board, as shown in the bottom left photo below. Once I've marked all high spots with a pencil, it's time to plane.

I set my scrub plane's blade so that it protrudes about 1/16 in. beyond the sole. I plane diagonally across the grain in parallel strokes, removing wood quickly and minimizing the chance of severe tearout, which would most likely occur if I went with the grain. Then I go back and plane on the opposing diagonal until I've covered the board (see the top left photo below). If I didn't have a scrub plane, I would begin flattening the board by using the jack plane in the exact same manner, but the process would just take longer.

When the work is more or less flat across, I switch to my bench plane, planing with the grain along the length of the board (see the photo at right below). I check for cup, bow, or twist every once in a while and again mark any high spots with a pencil. It's important to avoid planing low spots because they'd just become that much lower. If my plane starts to make dust instead of shavings, I resharpen the blade. I find that rubbing paraffin on the sole of a metal plane cuts down on friction tremendously.

My straightedge and winding sticks tell me when a board is flat, but it's evident, too, when I'm able to take long, lovely shavings over the full length of the board.

–P.K.

A SCRUB PLANE MAKES QUICK WORK of high spots on the rough board. Korn planes across the grain diagonally to prevent tearout and then planes on the opposing diagonal, removing the ridges created with the first passes. A bench plane is the next step.

SIGHTING ALONG WINDING STICKS tells the author that the board doesn't twist. He also uses a long straightedge to check flatness along and across the board's length.

TAKING A BENCH PLANE and planing with the grain, Korn takes out the scrub plane's marks and smooths the face of the board flat, readying it for the thickness planer and further milling.

WHEN JOINTING AN EDGE, choose whichever edge looks as though it will be less prone to tearout. Make sure the fence is square to the jointer beds and that the face against the fence is snug up against it. As when face jointing, transfer downward pressure from the infeed to the outfeed table as the board passes the cutterhead.

hand tools with no sacrifice in quality. Generally, I joint the face of a board and square a perpendicular edge (steps 1 and 3) on the jointer, plane the board to thickness (step 2) with a planer, and rip the board to width and crosscut it to length on the tablesaw (steps 4, 5, 6). If I have to flatten the face of a board wider than the 8-in. capacity of my jointer, I'll use handplanes for that step (see the sidebar on the facing page), but I still use machines for the rest of the sequence. If I didn't want to have to flatten a wide board with handplanes, I could rip the board in half, mill each half foursquare, and then glue the pieces back together. Handplaning avoids the extra glueline, and it's also one of the great pleasures of working with wood. There's nothing quite like unveiling a board's beauty with a well-tuned handplane, shaving by sinuous shaving.

Flattening the First Face

The setting of a jointer is critical to its performance. The outfeed table should be set at the highest point of the knives' rotation. A slightly high outfeed table will cause a board to become convex along its length. A low outfeed table causes snipe—the rear end of a board drops as it leaves the support of the infeed table, making the last few inches thinner.

The height of the jointer's infeed table determines the amount of wood removed with each pass. Take thin passes to reduce the possibility of tearout as well as wear on the machine. I never take off more than 1⁄16 in. per pass.

Whenever you joint or plane wood, you should check grain direction and ensure that you cut with the grain to avoid tearout. Where the grain is contrary, feed wood slowly and steadily and be sure the knives are sharp. Always use a push stick when using the jointer to flatten a face, so your fingers don't pass right over the cutterhead (see the bottom photo on p. 5). If a board is cupped and/or bowed—as most are—joint the concave side so that the board doesn't rock.

Making the Second Face Parallel to the First

A thickness planer's infeed roller, outfeed roller, pressure bar, and knives should be set according to the machine's manual. Also, the knives must be sharp and the table parallel to the cutterhead. I vary the setting of the table rollers in the planer's bed to suit the occasion. If I have hundreds of board feet to plane, I'll raise the table rollers a bit above the bed to help the lumber along. This causes a bit of snipe at the boards' ends as they are lifted by the table rollers, but I live with it because of the time and effort saved. When I want finer, more accurately machined stock, I lower the table rollers beneath the surface of the bed, which I keep waxed to help the boards slide along.

Snipe can occur even with the table rollers lowered. On a planer with an adjustable table and a fixed head, the table may be rocking, in which case the gibs that hold the table in place need to be tightened. On a planer with a fixed table and an adjustable head, the head assembly may need to be tightened in place. Check the manual for your particular machine if you have a problem.

To prevent tearout, thickness plane with the grain as much as possible. If a board has very squirrelly grain that has a tendency to tear out, feed the board through as slowly as your planer allows, and take thin passes. Never plane more than 1/16 in. at a time, in any case, to avoid stress on the machine.

Removing wood from the surface of a board will often upset its internal stress equilibrium and cause the board to warp. To maximize stability and flatness, I often stop planing when a board is between 1/16 in. and 1/8 in. from final thickness and let it readjust itself overnight. The next day, I reflatten one face with the jointer or handplane and take the board to final thickness with my planer.

Squaring the First Edge

Unless there are other considerations, the first edge I square is the one that can best be cut with the grain (see the photo on p. 7). Here is where the advantage of flattening both faces before truing an edge becomes apparent because I can now choose either face to run against the jointer fence. I always check the fence for square before jointing any wood, but I usually also make a test pass and check the board with a try square.

Ripping to Width

A rip blade in the tablesaw works well for ripping stock to width, but I prefer a combination blade when preparing stock so I won't have to change blades to crosscut the ends. Set the blade square to the table and just a tooth's height above the wood for safety. Then set your fence for the exact width of your cut by measuring from the rip fence to any sawtooth that inclines toward the fence.

WHEN RIPPING ON THE TABLESAW, safety should be foremost in your mind. Use a guard, splitter, and push stick, and make sure you stay out from behind the board you're ripping: Kickback happens faster than you can react to it.

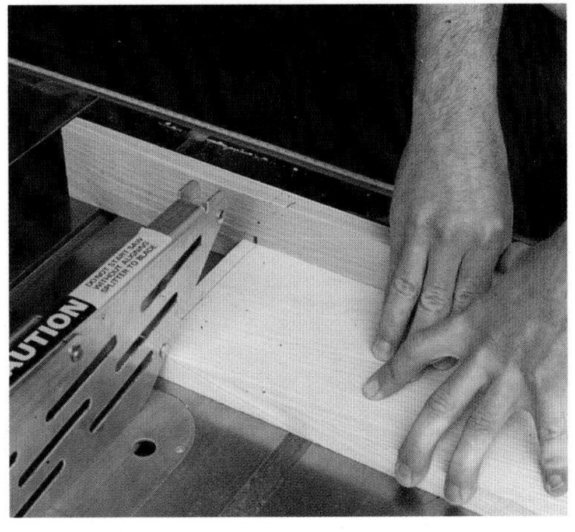

CROSSCUTTING ONE END SQUARE and the other end both square and to length can be done on the tablesaw with either a sliding table or a miter gauge with a wooden extension fence (left). The author lines up his cutoff mark with the inside edge of the sawkerf in the miter fence, ensuring an accurate cut.

There are two schools of thought about rip fences. One holds that the fence should be exactly parallel to the sawblade. The other believes that the fence should tilt a hair away from the back of the blade as extra insurance against kickback. I used to believe it was best to angle the fence away from the blade, but now I try to get the fence exactly parallel. Whichever you believe, just be sure the far end of your rip fence doesn't angle in toward the blade. At best, the wood will pinch and burn from friction; at worst, stock will catch and fly back at you faster than you can react.

Tablesaws are inherently dangerous, so here are some oft-repeated tips to take to heart: When ripping, keep the board firmly against the fence at all times, and push it with a smooth, steady motion (see the photo on the facing page). Never let go of a board until it is entirely past the blade. Use a push stick rather than pass your hand by the blade. Don't hold wood with a hand placed beyond the blade—your hand could be drawn back through the blade by kickback. Avoid standing directly behind the board being cut, and make sure no one else is in the path of potential kickback either.

Crosscutting the Ends Square and to Length

After stock has been flattened on both sides, jointed on one edge, and ripped to the designated width, it's time to cut the board to length. There are a number of tools with which you can crosscut. I prefer a tablesaw equipped with a combination blade. Crosscutting on the tablesaw is done with the aid of a sliding table or a miter gauge with an add-on wooden extension fence (see the left photo above). Never crosscut with the end of a board against the rip fence.

After the first end is cut square, you can either measure out the desired length on the stock and pencil a cutoff mark, or you can attach a stop to the fence of your sliding table or miter guide at the desired distance from the sawblade. The quickest way to cut to a pencil line is to align it with the edge of the kerf the blade has left in the fence—as long as you always use the same blade (see the right photo above).

From the moment wood is milled, movement should be minimized by careful handling. To promote even exposure to air, I either leave boards on edge or stack them horizontally with spacers between them. I also keep wood away from direct sunlight and any other heat source that could affect one side of a board more than the other. I also try to cut all joinery right away while the wood is as square and straight as it will ever be.

PETER KORN is the Executive Director of the Center for Furniture Craftsmanship in Rockport, Maine.

From Rough to Finish

BY GARY ROGOWSKI

I have paid lumberyards good money for some nasty-looking hardwood. Sometimes you just have to take what you can find, even if the stock has defects. But I do have some faith in the power of machines. Planks that look like they were pried off the hull of a beached boat can be made silky smooth and straight as an arrow with the push of a button.

Well, almost. You can't blindly shove stock into the maw of a groaning machine and extract perfect boards. If you repeatedly pass the face of a twisted board across a jointer and don't apply proper pressure to the opposing corners, you'll end up with one big shingle—skinny on one edge and fat on the other.

The first step in milling is looking, not machining. Examine your stock, and identify problems such as bow, check, cup, and twist. Different defects call for specific milling strategies. But even when you're careful to identify problems, surprises sometimes arise. Recently, while planing a plank of what looked like clear sycamore, I noticed a sudden color change in the machined face.

USING MACHINES TO REMOVE CUP, crook, twist, and other defects from lumber.

Warped View of Lumber

As wood dries and ages, strange things can happen to it, even under the best of conditions. Identifying the problem is the first step in milling stock efficiently.

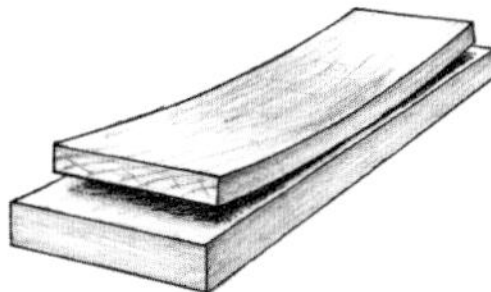

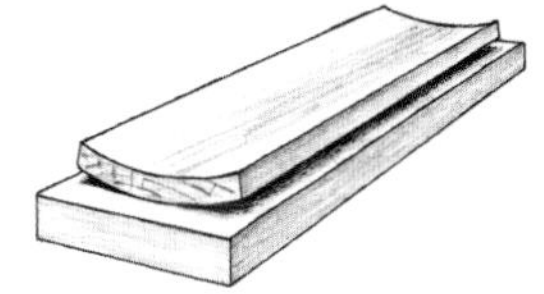

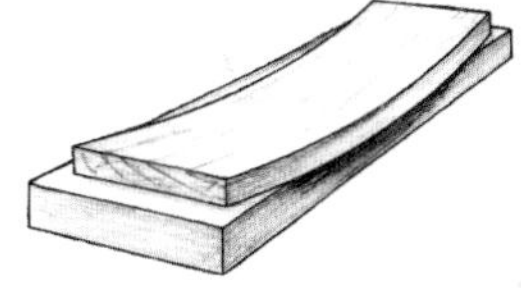

Bow
Bow occurs along the length of a board on the face side. If the bow is severe, it is best to cut the board into smaller sections before machining.

Cup
Cup occurs across the face of a board. If lumber is badly cupped, rip it into narrower sections; you'll end up with thicker stock after face-jointing and planing.

Twist
Lumber with a slight twist will give you fits if not removed prior to cutting joints or gluing panels together. If it's severely twisted, cut the lumber into shorter sections for better yield.

Crook
Crook is a bow along the edge. You'll end up with waste along both edges when ripping it straight and parallel.

Checking
Checks may occur throughout lumber, but they are most commonly found at the ends of a board, the result of too rapid drying.

CHECKS CAN BE FOUND ANYWHERE. **Though they are most common at the ends of boards, checks may also occur in the middle of a board (below left). In the case of internal checks, the problem may not be obvious until a board is crosscut (below right).**

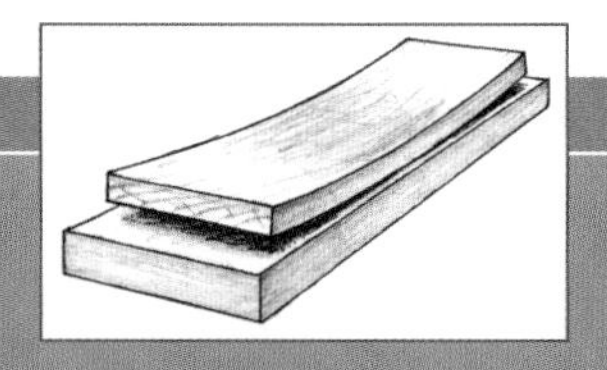

Bowed Lumber

TO DETERMINE WHETHER A BOARD IS BOWED, SIGHT DOWN ONE EDGE. Bowed boards are best used for shorter pieces of a project. Mark sections using a pencil while eyeballing the amount of bow. Next crosscut the board into shorter sections; then joint them flat, placing the stock bow side down on the jointer table (below).

SET THE MACHINE TO TAKE SHALLOW CUTS, ABOUT 1/32 IN., FOR ALL FACE-JOINTING. The jointer will remove material at the ends first (right). Be careful not to exert too much pressure on the board, or you may temporarily press the bow out, resulting in a board that planes unevenly and isn't flat.

I took a closer look. Smack in the middle of the discoloration was a chunk of buckshot. Fortunately, the soft lead didn't damage the planer's knives.

Although it may seem like more work, I prepare stock in two steps: rough milling and finish milling. First I pick through the stock and decide what boards to use for which parts of a project. Next I crosscut the pieces 1 in. oversize in length, rip them on the bandsaw, leaving them 1/8 in. over in width. Then I joint and plane the stock, leaving everything 1/8 in. over in thickness. When rough milling, I concentrate on the serious defects and don't worry too much about getting perfectly square edges yet.

Then I sticker the stock for a few days to allow any hidden stresses in the wood to reveal themselves.

Wood that's been sitting in a rack may hold hidden surprises that show up after milling. After letting the stock settle down, I'll do the final milling—getting stock square and cutting it to the final dimensions. By then, the stock is usually pretty stable and less likely to play tricks on me.

The defects found in lumber are often a result of what happened to the wood before you bought it. As wood dries, even under ideal conditions, it suffers some degradation. Improper drying—too fast, too slow, improper stickering, and other mistakes—can play havoc with wood. Here are some of the more common problems and how to solve them.

Use Bowed Stock for Short Pieces

Bowing describes a board bent along its length on the face side. Bowing isn't too great a problem if you need short pieces. You can dress the face of a short bowed plank until flat. But for long tabletops, where you need the thickness, bowing can cause problems. One end or both will wind up too thin after repeated passes over a jointer. When a project calls for long pieces, and the lumber is bowed, select stock thicker than needed to allow for waste.

Face-joint bowed stock concave side down across the jointer. Severely bowed stock may catch on the outfeed table as soon as it passes over the cutterhead. If it does, lift the board onto the outfeed table. Then push the stock through. Repeat until the board no longer hangs up. Alternatively, you can joint enough of a flat onto the rear of the board until the front end no longer catches. Don't exert too much pressure, or you may temporarily press the bow out. I set my jointer to take very light passes—about 1/32 in.—for all operations, even on

Cupped Lumber

A STRAIGHTEDGE PLACED ACROSS THE FACE OF A BOARD WILL INDICATE THE AMOUNT OF CUP. Remove this flaw by placing the board cup side down on the jointer (left). The machine will take off material at the outside edges first (below). If the cup is severe, you may end up with stock that's too thin. To avoid this, rip the stock into narrower strips on a bandsaw before face-jointing.

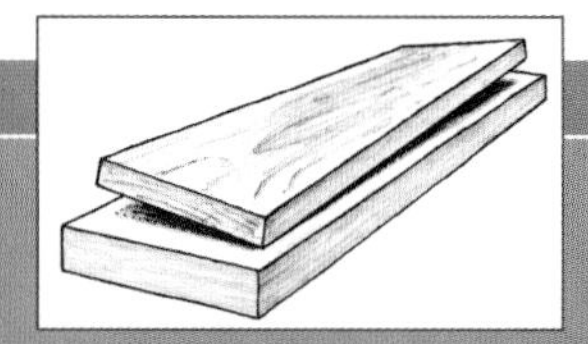

Twisted Lumber

USE WINDING STICKS TO CHECK LUMBER FOR TWIST. Lay the sticks across the board at opposite ends, and sight down the board. If the sticks aren't parallel, the lumber suffers from twist.

REMOVING TWIST ON THE JOINTER. This is accomplished by taking a diagonal cut across the face of a board. Begin by pressing the lead high corner flat to the table. Gradually transfer pressure to the trailing high corner as the board passes across the cutterhead. Don't let the board rock onto the low corners, or you will remove material where you don't want to.

Winding Sticks Help Identify Twist

Winding sticks are simple but accurate tools that help you spot twist in lumber. They're handy for truing up other surfaces as well, such as your bench or jointer tables. Mill up two sticks about 1 in. by 2 in. by 24 in. Make sure that the wood is dry, knot-free, and straight and that the two pieces come out the same size. Mark along the edge of one stick using a dark marking pen, or for fancier sticks, make an inlay of darker wood.

To use the sticks, place one on each end of a board. Move away, and then hunker down and sight from the top edge of the near stick to the top edge of the far stick. If the two sticks are parallel to one another, the board is flat. If the sticks are tilted with respect to one another, the board is twisted. To remove twist, the board is face-jointed, and the high corners are removed first. –G.R.

rough stock. It's easier on the machine and easier on you. A bigger bite means more vibration, which will reduce your ability to feed stock smoothly. I also use a push stick on the back edge of a board.

Jointing a High Spot

A board with a hump on one edge requires a balancing act to get a true edge. Place the board on the infeed table of the jointer, and put your weight onto the trailing end of the board. This will lift the lead end of the board as it passes over the cutterhead. Slide the board along until it just starts to cut the hump. Then transfer all your pressure to the outfeed section of the board, which will lift the rear portion off the infeed table. Repeat until the stock doesn't rock and material has been removed across the entire face.

If your lumber has wild or swirling grain, often found near small knots, use a damp rag to lightly moisten the wood fibers before cutting. Take shallow passes when jointing or planing, removing less than $\frac{1}{32}$ in. at a time. This will help avoid

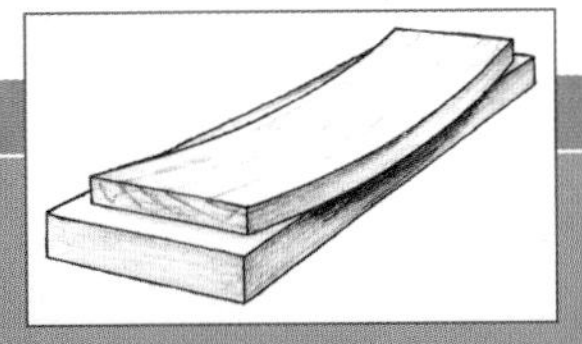

Crooked Lumber

CROOK CAN BE SAFELY REMOVED USING A BANDSAW Use a batten or any straightedge the length of the stock, and mark the area to be removed with a pencil. The author prefers using a bandsaw for all rough-ripping because there's no chance of kickback.

tearout. The same method works well for lumber with wild grain, such as curly maple.

Taking the Cup out of a Board

A moisture imbalance between two faces will cause a board to pull itself into a cupped shape. The side with more moisture will expand at a greater rate and become convex; the drier side will shrink and become concave. You can spot cupping by sighting across a board or by holding a straightedge across its face.

To flatten a cupped board, place the concave side face down on the jointer. Take light passes until the entire face has been touched by the cutter. Flatten the convex side by running the board through the planer, humped side facing the cutterhead, after face-jointing. When setting the depth of cut on your planer, reference it off the highest part of the cup.

Most Rough Lumber Has Checks in the End Grain

End-checking or cracking is common in all lumber. As wood dries, moisture escapes faster from end grain than from the face or edge. That's why it's important to paint the ends of green lumber before drying it, which will help equalize the rate of shrinkage throughout the boards. Nevertheless, end-checking occurs frequently. When buying stock, factor in the loss of a few inches of length.

Although less common, lumber may also check along its surface, far away from the ends. This occurs more frequently in certain species such as oak. These checks tend to be narrow—⅛ in. or less. Lumber that has been dried too quickly may develop severe internal splits. These splits may be in the form of interlinked cracks called honeycombs or one large massive crack running the entire length of a board. You can sometimes spot a honeycombed section by looking for a bulge on the face of a board.

There are various methods for dealing with checked lumber. For a simple solution, cut off the afflicted sections, and use them for firewood. Some woodworkers celebrate these natural flaws by filling them with colored epoxy resin or cutting a butterfly key to stabilize the crack.

Gone with the Wind

A twisted board is the most sinister of defects. Slight twist—also commonly referred to as wind—may go unnoticed until you begin face-jointing a board and realize too late that you've created a taper. When you try to correct it by more face-jointing, you may end up with stock that's too thin at one end.

Check for twist by sighting down one end of a board to the other. If one corner appears higher than another, the board is in a twist. Tools called winding sticks are a foolproof way to help you spot twist (see the sidebar on p. 14). A flat surface such as a workbench also can be used as a tool to look for wind. Place the lumber face down, and push on the adjoining corners. If the board rocks, it's twisted.

If lumber has other faults besides twist, such as bow or cup, deal with the twist first. Place the board on the infeed table of the jointer, and press down on the low corners. Exert greater pressure at the front of the board at the beginning of the cut; then transfer pressure to the rear as it approaches the cutterhead. The board will be cut across a diagonal line from one high corner to another. Repeat until the board is flat.

Remove Crook with a Saw

Think of crook as a bow along the edge of a board. The same problems encountered when jointing bowed lumber may occur with crooked boards. First crosscut the stock into approximate lengths needed for a project, then rip the boards slightly oversize using a bandsaw. This will make it easier to joint an edge straight without wasting a lot of wood.

As with bowed wood, if you're having a problem with the stock catching on the edge of the outfeed table, place the leading edge of the board on the outfeed table, just past the cutterhead, then push it through. Continue until the board no longer catches, jointing it in the usual way.

GARY ROGOWSKI is a contributing editor to *Fine Woodworking*. He runs The Northwest Woodworking Studio, a school in Portland, Oregon, and is the author of *The Complete Illustrated Guide to Joinery* (The Taunton Press, 2001).

From Rough to Ready

BY ROGER A. SKIPPER

I'm a woodworker and a writer. I also enjoy eating regularly, so I have found other ways to supplement my income. A few years ago, I built a 3,000-bd.-ft. lumber kiln. Because many of my clients are basement woodworkers who have no practical way to turn rough lumber into finished stock, I am often asked to dress their stock as well as dry it. No problem. No problem, that is, if you have a 16-in. jointer and a ripsaw. I don't.

When forced to dress 3,000 ft. of lumber at a clip, from wide and twisted material to long and crooked, I found that standard small-shop methods were unwieldy and slow. So I developed methods that streamline the process and also work well for a small shop on a shoestring budget. And these procedures are valuable for dressing any large amount of lumber, from 100 bd. ft. to

PLANING

EFFICIENT STOCK PREPARATION WITHOUT A 16-IN. JOINTER AND A RIPSAW. The author planes first, moving easily around an ergonomic workstation. A nearby radial-arm saw chops off problem areas, infeed and outfeed piles are readily accessible, and boards that are twisted or too thick or thin are stacked close by to be dealt with later.

Timesaving Go/No-Go Gauge

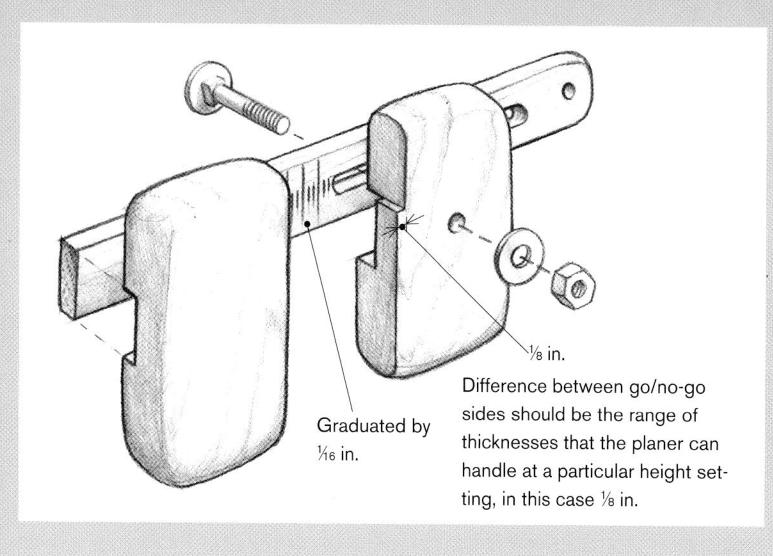

A SMALL GO/NO-GO GAUGE identifies boards that fall within the planer's ⅛-in. cut. Many sawmills don't deliver boards of uniform thickness. Valuable time is wasted when a tapered board binds in the planer or a thin board won't feed. Too-thick and too-thin boards are put aside for later passes.

thousands. The system combines efficient materials-handling, a few shopmade sleds and accessories, and a logical order of operations. Here's how I do it.

Plane First

I tackle the planing operations first. Most of the lumber I dry has been cut on a portable mill. My biggest problems are the overall range of thicknesses and single boards with tapering thicknesses. Because my planer's maximum bite is about ⅛ in., it doesn't take much taper to stall a board midway through the cut. It is also frustrating to watch your valuable time trickle away as a board too thin to reach the knives plods to a halt again and again. I sort the lumber as it comes rather than constantly adjust the planer. Find an average thickness of the lumber you are going to plane, and set your planer to take a medium cut from that measurement.

Go/No-Go Gauge A two-sided thickness gauge will save you countless hours and aggravation. Measure the thickest part of the plank. If it's between the gauges, send it through the planer. If not, throw it back into either the too-thick pile or the too-thin stack. Also make a separate pile of twisted or bowed lumber for later processing. As you feed the stock, flip any crowned lumber so that the concave side is down. This ensures that the edges will be of similar thickness and that the thin spot will end up in the center of the board. If it is fed in with the concave side up, the board will be unstable, and the pressure of the rollers will dominate on one side, with the planer biting deeply into the other edge. This often produces a board that is unusable for half its width.

Bark edges, common on rough lumber, often contain dirt and other debris that is bad for the knives, so I keep a drawknife handy to strip away bark. I constantly watch for hazards, such as loose knots or rotten areas, and cut them away with a small radial-arm saw. Broken ends that come to a point can get wedged under the next piece of stock being fed through the planer and also must be cut away.

I am a one-man band, so I handle the outfeed side, too. I stack the planed lumber on rolling carts, close to the outfeed table.

At the completion of the pass, I move the stack of lumber back to the starting point and offbear onto another set of wheels. With my radial-arm saw nearby, and my infeed and outfeed stacks on both sides, I can work through my piles without having to move around much and without being out of reach of the planer's off switch.

How to Plane Twisted and Bowed Lumber

Twisted and bowed stock will not yield full-thickness material. The more pronounced the bow or twist and the longer the length of the piece, the thinner the finished material. I cut these planks into 4-ft. pieces, sacrificing length to maintain thickness. Narrow lumber is simply leveled on one side on my 6-in. jointer, then planed to thickness. But all twisted planks wider than 6 in. are handled on the planer, with the help of the following accessories.

Leveling Strips for Heavy Stock Wider lumber, if only moderately twisted and if heavy enough to resist the pressure of the feed rollers, can be milled flat using leveling strips (see the photos on p. 20). Lay the plank on a workbench or other flat surface, shim under the high corners to level the board overall, and lay a couple of ¾-in.-thick strips along the sides. Attach the strips with 1¼-in. screws. This gives a level surface to pass over the planer bed and will result in a flat surface on top. Then remove the strips, turn over the piece, and plane it just enough to clean the board. The screw holes go in only about ½ in. on each side and can be ripped away.

Bowed lumber can be handled using the same method, with the concave side down.

Warning: Don't yield to the temptation to insert a row of screws in each strip and use only the necessary ones, allowing the others to protrude. These screws can come loose and become missiles and blade-destroying foreign matter. Remove all screws not in use.

Although this process is not a quick one, it allows the salvage of valuable lumber that would be wasted without access to a wide jointer. Twisted lumber is often highly figured and worth saving.

A Sled for Thinner or Severely Twisted Stock I developed this leveling sled to

Planing Thick, Twisted Boards

THIS BOARD IS WORTH SAVING but is too wide for the author's jointer. The author uses leveling strips for thick boards that can withstand the pressure of the planer's feed rollers and aren't twisted too severely to line up with the 1¼-in.-high strips.

THE LEVELING-STRIPS METHOD for twisted lumber. The plank is placed on a flat surface, and shims are placed under the high corners to even them out. Hardwood strips (¾ in. thick) are then screwed onto the sides, using 1¼-in. multipurpose screws. The strips offer three possible hole positions at each location.

THE STRIPS PREVENT THE BOARD from rocking as it passes through the planer, yielding one flat side. Then the strips are removed, and the board is flipped over and planed to a uniform thickness. The screw holes are ripped off the edges later.

support severely twisted or thin stock that would be compressed by the planer's feed rollers (see the photos and drawing on the facing page). The sled consists of a stiff table with adjustable leveling supports placed every 6 in. I crosscut twisted lumber to 4 ft. or less, so I built a 4-ft.-long sled.

Lay down the board with one side and one end butted against the plywood fences on the edges of the sled, then adjust the supports to fit it underneath. The twisted lumber must be crosscut to a 6-in. increment so that the front end of the board lands on one of the adjustable supports; otherwise, the planer's feed rollers will force down the front end of the board and snap up the back end.

Run the sled through the planer until the top surface of the board is flat. Then remove the lumber from the sled, flip it over, and plane it to a uniform thickness.

When all of your stock is flat, you are ready for ripping.

Sled Ensures a Straight Edge

My idea for a shopmade ripping sled came from observing the operation of a small rotary mill. Logs are loaded onto a sliding carriage that passes by a fixed blade. The straight sliding action is the key to producing the straight ripped edge.

My tablesaw operation works on the same principle. I made a runner to slide in the T-slot of my saw table and attached it to the bottom of a carriage board. The front of the lumber is held in place by jamming the end into several sharp multipurpose screws. I hold the back of the board in place by hand as I push it past the blade. I originally put an elaborate T-slot on the top of this sliding carriage, with a sliding clamp to hold down the lumber. But I found this clamp to be overkill.

Building the Ripping Sled My sled had to be 16 ft. long; yours should be as long as the longest stock you use. Almost any length will require rolling infeed and outfeed support. I use 4/4 pine for the carriage board. Only the miter-slot runner will ride on the support rollers, but the carriage board is stabilized by the saw table.

The sled moves quickly and could jump out of the miter slot and into the spinning blade in an instant. To hold down the table during use, I recommend attaching a wider

Planing Thin or Severely Twisted Boards

THE PLANING SLED IS TIME-CONSUMING to set up, but some boards are worth it. Butt the board against one side and one end of the sled and snug up and hand-tighten the end supports. Then snug up the rest of the supports and hand-tighten their fasteners. Note: This sled only accepts stock that is 4 ft. long or less, and the stock must be cut to a 6-in. increment in length so that its end is supported.

REMOVE THE BOARD TEMPORARILY. And tighten down all of the bolts.

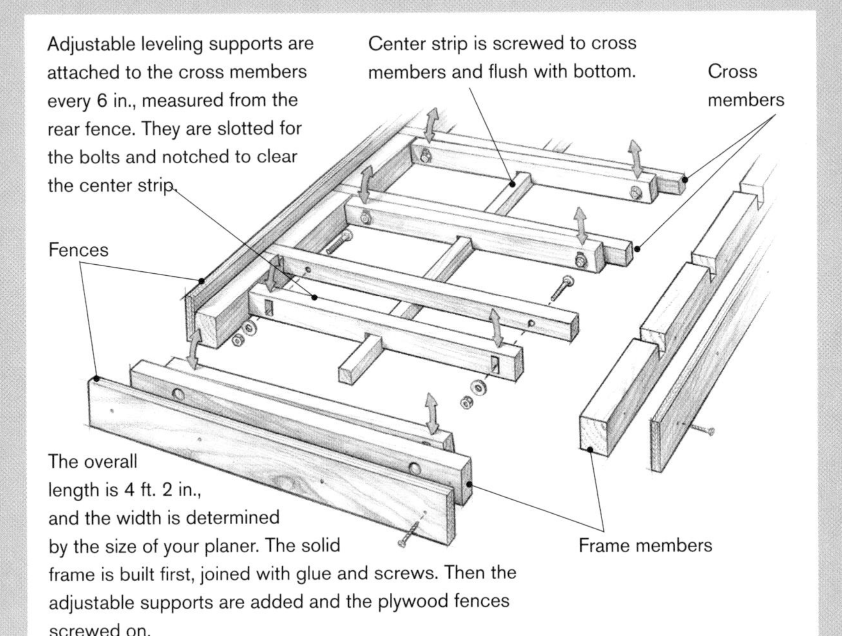

The overall length is 4 ft. 2 in., and the width is determined by the size of your planer. The solid frame is built first, joined with glue and screws. Then the adjustable supports are added and the plywood fences screwed on.

THE ENTIRE SLED GOES THROUGH THE PLANER. Multiple passes will probably be necessary to produce a flat side.

strip to the bottom of the runner, to fit into the wide part of the T-shaped miter slot.

A T-shaped hardwood runner will stand up to home-shop use. But I use a more heavy-duty version built up from three ⅛-in.-thick aluminum strips, with ⅞-in. fender washers fixed loosely to the bottom. The bottom strip is interrupted for each washer, allowing it to sit flush. The washers are held only loosely by screws to allow them to wiggle around the sawdust that builds up in the miter slot.

Attach the runner to the carriage board so that a bit of the board will be trimmed off the first time past the blade. The edge of the sled will indicate exactly where the sawblade passes. In use, you will be able to see and feel the overhanging portion of a board that will be cut off, and position the board for optimal ripping.

Ripping a Straight Edge on Crooked Boards

THE AUTHOR'S SHOPMADE RIPPING SLED rides in the miter slot on the tablesaw. Hand pressure at the back of the board and screw tips protruding at the front of the sled secure the workpiece as it moves past the blade. Again, infeed and outfeed piles are easily accessible and placed on rolling carts.

ENGAGE THE PIVOTING STOP BEFORE loading a board. The small wood stop will keep the sled from sliding forward toward the blade while the board is being pressed onto the screw tips. The stop is simply trapped against the edge of the saw table and the front rail of the fence system. When the board is loaded, the sled is drawn backward to release the stop, which will then pivot out of the way when the sled moves forward. Screws left protruding on each side keep the stop from flipping over the top in use.

THE EDGE OF THE SLED IS TRIMMED FLUSH with the blade, so it is easy to tell how much material will be removed. You can see and feel the overhang and thus optimize the amount being ripped away.

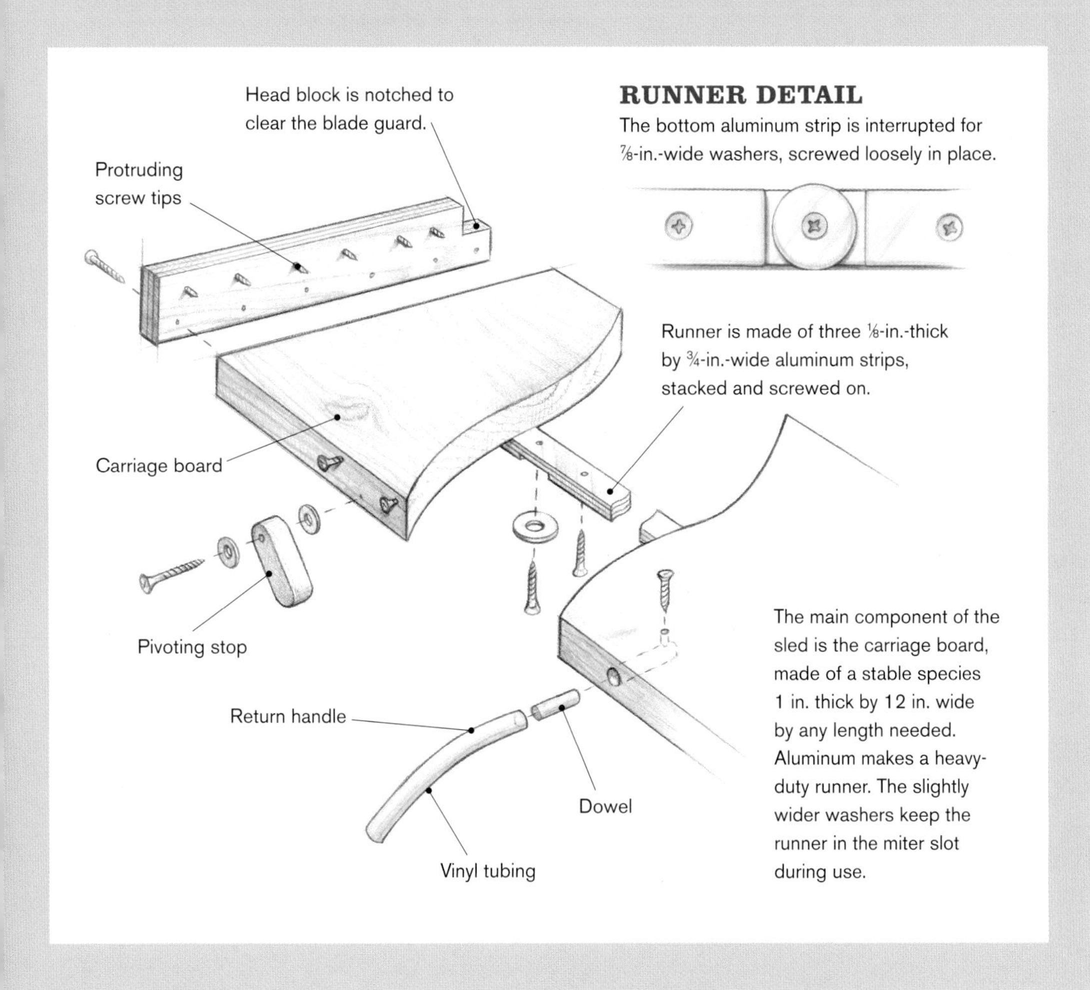

Rolling Carts for Moving Lumber

The heart of the rolling cart is a pair of heavy-duty swiveling casters. I purchased mine from the Northern Tool & Equipment℠ catalog. For easier rolling, use wheels that are at least 6 in. dia., and size the weight capacity to reflect the loads you intend to carry. I got my heavy-duty casters for a little more than $20 each. I opted for steel wheels, which roll well on my concrete floor. If you are working on a wood floor, as is the case with the local lumber company that is using my system, rubber wheels are available for about the same price.

Use a section of 2x10 or 2x12, and screw plywood skins onto the top and bottom. Attach the casters with carriage bolts.

Don't think that you can use this method in your parking lot. It is a smooth-floor system only.

Also, be aware of the concentrated weight exerted by each caster, and make sure your floor can handle the load.

WHEEL PAIRS ARE EASY TO STORE. Two-caster assemblies take up far less space than a complete four-caster frame would.

Snap a line on the carriage board for aligning and attaching the runner. Because a chalkline is slightly fuzzy, and it is critical that this runner be straight, strike a sharp pencil line down the center of the chalkline, using a long straightedge. Fasten the runner to the board, countersinking the screw heads. When the runner is fastened, adjust your roller tables so that the bottom of this runner, not the bottom of the carriage, passes smoothly from the saw table to roller table. Good adjustment here will allow easy rolling later. The table will probably slide hard at first but will loosen up with use. A little paraffin wax helps.

All that remains is to put in the head block, a piece of plywood with screw tips protruding on the inside; a pivoting stop to keep the table from drifting into the blade while a board is being loaded; and return handles. The stop is a simple piece of Plexiglas® or wood with a couple of screws added to keep it from flipping over the top. The return handles are pieces of vinyl tubing used to pull the table back to its infeed position after a pass. A series of these handles allows one to be handy wherever you end up. I installed rigid return handles on my first version of this sled, but a few painful encounters with these convinced me that flexible handles were better.

Using the Ripping Sled Do yourself a favor and get a blade designed for ripping. The ease and speed of cut will make the purchase worthwhile.

Latch the sled in the rearward position. Lay a board to be straightened on the carriage, allowing whatever you wish to cut away to hang over the edge. Push the board forward into the screws in the head block. Pull the sled backward a couple of inches to release the pivoting stop, then smoothly feed both the stock and sled through the saw as you hold down the rear of the board against the carriage. Push the sled past the blade a safe margin before removing the stock.

Severely crooked boards can be cut in half lengthwise, reducing the crook by a factor of two.

These stock-preparation methods aren't costly or elaborate, but they work very well. They'll get your next major project off the ground more quickly.

ROGER A. SKIPPER is a lumber processor, construction consultant, and instrument maker in Oakland, Maryland.

Choosing a Tablesaw

BY ROBERT M. VAUGHAN

In many small shops, no tool is more important than the tablesaw. Whether you're looking for your first 10-in. tablesaw or satisfying an upgrade itch, you're more likely to get the saw you want if you start by defining your needs—and your budget. By knowing where to look and what to look for, you'll stand a good chance of buying the right one.

For cutting ¾-in.-thick pine, almost any saw will do. But if you're ripping 2-in. hardwood, the more power you can get, the better. So the first step is to analyze the type of material you usually work with.

If you work with a lot of plywood, bolt-on table wings and an outfeed table will be very helpful. And a hefty saw with a long, sturdy rip fence will make working with heavy panels easier.

Also, consider your skill level. An accomplished woodworker can do good work on almost any machine. If something goes wrong, a beginner may have trouble figuring out if the problems are due to a

THERE'S A SAW for every budget, as shown in this display of saws. To get the right saw, define your needs first. After that, it's easy to find a saw with the right combination of power, quality, and precision.

Starting out on a good machine will shorten the woodworking learning curve.

lack of craftsmanship or the machine. Starting out on a good machine shortens the woodworking learning curve, assuming the saw has been set up correctly.

Shop size also will factor into your decision. Do you have space for a machine in the middle of the floor? Or will you need to push it off to the side when not in use? For the truly tiny shop, a small, lightweight machine that can be stored under a workbench might be just the ticket.

Where to Look for New Machines

I strongly recommend seeing the machine before making any buying decisions. Hardware stores and home centers are good places to find medium-to-low-range machines. Higher-quality machines are sometimes found at home centers and large hardware stores but more likely at industrial distributors. The biggest problem with looking at any new machine on a showroom floor is that you may not be able to try it out.

One way to get around this problem is to finagle an invitation to someone's shop. A manufacturer or distributor may know a customer who is willing to show off his equipment. Once in the shop, you should be able to see firsthand where the machine throws dust, how loud it is, and how convenient the controls and switches are. Be sure to ask the owner about the saw's weak points as well as its strengths.

Weigh Mail Orders Carefully

I'd rather buy a new tool from a local dealer than from a mail-order company. For one thing, I get to see the tool before I buy it. I

TOP-QUALITY SAWS cost between $1,500 and $2,000. These saws, what the author describes as Class A machines, have solid cast-iron tables and extension wings and a full-length fence. These heavy-duty, smooth-running machines form the backbone of many small professional and serious amateur woodworking shops.

IMPORTS MIMIC HIGH-END MACHINES. **Imported imitations of more expensive machines offer woodworkers many big-machine advantages for less than $1,000, but expect some compromises in quality.**

also want to help keep local retailers in business. A local dealer may help set up a new tool or at least provide advice if I run into trouble. And if I don't like the tool, it's a whole lot easier dealing with a local retailer than a mail-order company hundreds or thousands of miles away. A mail-order house will sell you a machine, but the company sure can't service it.

But there are advantages to buying mail order, too. Mail-order prices are often (but not always) lower. Your choice of brands is likely to be much wider. If you shop locally, you may have to settle for a tool that isn't your first choice. With so many mail-order houses to choose from, buying by mail means you get exactly what you want. If you do shop by mail, though, make sure parts will be available.

How to Look at a Tablesaw

A pocket flashlight, a nickel (yes, a nickel), a tape measure, and a piece of string are helpful when looking at tablesaws. The flashlight will help you investigate the guts of the machine. Check out the gauge of the sheet metal, thickness of the cast iron (you don't want thin, tinny parts), and overall heft of the machine. Shake anything you can grab: the motor, the drive belt, the trunnions, even the blade. (Make sure the saw's unplugged, and take care not to cut yourself on the sharp teeth.) Parts that rattle easily may fall off later or indicate the saw isn't well-made. Listen to the machine when it's running. Are there vibrations, rattles, or other suspicious sounds?

Vibration in a tablesaw can cause ragged cuts and can contribute to operator fatigue. If you are able to test the saw before buying it, check for vibration by balancing that nickel on the saw's table, with the machine running but no blade attached. If the nickel doesn't fall over, good. If the nickel won't stay up, look at a different saw.

Measure the machine's footprint, including anything that sticks out, such as switch boxes or motors, to see if the saw will fit your allocated space. Check to see that the miter-gauge slots are parallel. Pay particular attention to this detail on less expensive machines.

The string is used as a gross check of the tabletop's flatness. Stretch the string across the top in several places, and look for any dips or humps.

What You Get for Your Money

To better define what a buyer might expect to get for his money, I've divided tablesaws into five different classes based on price. Class A saws range from $1,500 to $2,000 and include saws such as Delta's® Unisaw®, the Powermatic® 66 and the General® 350, as shown in the photo on the facing page. These 10-in. saws set the standards by which all other saws are judged. The Class B saws cost between $1,000 and $1,500 and are stripped versions of Class A saws. The $500 to $1,000 Class C saws are the contractor-type saws, as shown in the top left photo on p. 28, and the Taiwanese versions of Class A saws, as shown in the photo above. The $300 to $500 Class D saws are mostly lightweight saws or imported

CONTRACTOR'S SAWS are one of the most popular options for the home shop. Also found in many professional shops, the contractor-type saw usually sells for $500 to $1,000.

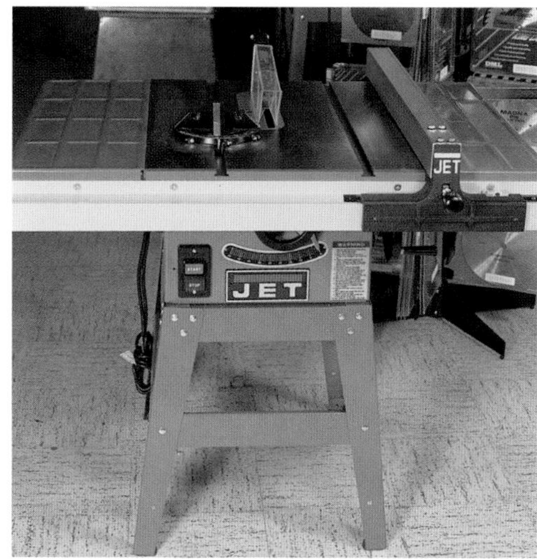

IMPORTED COPIES of the contractor-type saws appeal to occasional woodworkers (above). Saws like these are available for $300 to $500.

contractor saws (see the photo above right). Less than $300 buys hobby- or homeowner-grade machines.

Class A Tablesaws: $1,500 to $2,000

The strength and purity of the highly refined cast iron and steel used in these saws help make them the smoothest running and most durable 10-in. saws on the market. Standard equipment includes a $300 fence, a $300 motor, a $60 miter gauge, and often a $150 magnetic starter.

But the internal framework is the real strength of these machines, as shown in the photo on p. 30. The trunnion brackets are large and strong. The worm gears and machined teeth are large and made from high-quality metal. Hand wheels are large and easy to use. The enclosed base of a Class A saw houses the motor, which reduces the machine's footprint and makes dust collection a lot easier.

Service and tune-up on a Class A saw is easiest of all 10-in. tablesaws. By removing only three or four bolts, the table can be taken off to expose the internal workings of the saw for bearing or belt changes, lubrication, and cleaning. Parts are readily available and will continue to be available. And there are lots of aftermarket accessories made for these saws. Because of their desirability, these saws maintain a high resale value.

Though Class A saws are powerful, smooth, and accurate, they aren't perfect. Two-inch oak still bogs things down unless a rip blade is used. The power these saws use will require a separate circuit. Because they're so heavy, a roll-around base is helpful if the saw will be moved frequently. And, because of their power, kickbacks tend to be a little more forceful.

Class B Tablesaws: $1,000 to $1,500

Stripped-down versions of Class A saws, these saws often have lower-powered motors (1½ hp or 2 hp instead of 3 hp, for example), inexpensive electrical controls, or short fences. On the low end of this price range, you'll find some saws with Class C contractor-saw inner workings dressed up like Class A saws, complete with fence and accessories.

Parts and service for Class B tablesaws are the same as the Class A saws, and resale values, while not quite as good, are still relatively high.

Class C Tablesaws: $500 to $1,000 This is probably the most popular price range of tablesaws found in nonprofessional shops because of the balance between quality and price. At the upper end are the Taiwan-made copies of the Class A saws and at the lower range, the lighter saws made from high-quality cast iron and steel, such as the Delta contractor's saw.

I've heard mixed reviews from owners of the half-priced copies of Class A machines. Schools or professional shops probably won't be completely satisfied with a saw in this price range. Nonprofessional woodworkers don't seem to be as sensitive to the consequences of quality compromises because they don't use their saws as much. The problems usually start with the motor or motor controls.

The internal workings of these saws are similar to those in better grades but aren't up to the same standards, as the price clearly reflects. The fences on these machines seem to be satisfactory but, again, not of the same caliber as more expensive saws. Parts and service records are spotty, with some importers better than others. Predictably, the resale value of these Taiwanese machines isn't as strong as it is with more expensive saws.

WORKING PARTS **of mid-range contractor's saws are lighter than top-of-the-line saws and transmit more vibration. But the parts of these Class C saws are plenty strong due to quality cast iron and steel.**

The internal components of the contractor-type saws are somewhat lightweight, as shown in the photo below. But the full-sized table, combined with a low price, has made this saw a hobbyist's favorite for years.

The lightweight design, though plenty strong because of the quality of the metal, transmits more vibration than the Class A saws, particularly when using a dado set or a molding head. The motor is suspended out the back of the machine, and the long-drive belt contributes to the vibration. The suspended motor also may interfere with an outfeed table, and without an outfeed table, cutoffs drop on the motor and could get wedged in or damage some of the mechanical components.

The bottoms and backs of these saws are open, thus presenting a challenge when hooking up a dust collector. Motors, usually 1½ hp, and electricals on these machines are adequate.

The tabletop is full-sized, so most of the accessories that fit the Class A machines also will fit contractor-type saws. And parts are readily available. Servicing the machine is a bit of a pain because everything bolts to the bottom of the tabletop and requires flipping the machine upside down for some precision blade adjustments. The standard fence works quite well when adjusted properly. The resale value of these machines remains good.

Class D Tablesaws: $300 to $500 This group is the most popular for entry-level purchases. In this group are Taiwan-made copies of contractor-type saws and direct-drive, motorized saws.

The Taiwanese contractor-saw copies have all the inherent problems of the relatively lightweight contractor-type saws plus a few new problems. The motors and switches don't have the same longevity as those on the better-made machines. Many of the motors are advertised as totally enclosed, fan cooled (TEFC), but they aren't, as removal of the fan cover and fan quickly shows.

Bearings can be another problem with Taiwanese saws in general. Sawdust contamination of the shielded or open bearings found in most Taiwanese saws can result in premature bearing failure.

Another type of saw in this class are Class E saws with minor upgrades, such as better tables and fences. Expect a lot of soft aluminum extrusions on these models. Handwheels and handles are generally made of plastic. These parts are more susceptible to breakage than those on better-quality machines. The face surfaces of fences and miter gauges often are not as flat, straight or perpendicular to the table surface as they are on higher-quality saws.

Service is spotty, and many parts are difficult, if not impossible, to install, such as brushes and tiny cogged internal drive belts. Resale value on these machines is not as good as previous classes.

Class E Tablesaws: Less Than $300

These saws are the least suited for the beginning furnituremaker. The beginner often blames himself for imprecise work that is directly attributable to the saw. These saws are usually small, benchtop models that are light enough to be portable.

INTERNAL FRAMEWORK is the real strength of top-of-the-line saws. Heavy-duty trunnions, gears, and bearings make for easy adjustments and vibration-free operation in saws costing $1,500 to $2,000.

The tilt and raise mechanisms are usually lightweight, as shown in the photo at right, and get sticky and imprecise with the introduction of sawdust and use. The accessories usually are flimsy. These saws generally are considered disposable because of the expense and difficulty involved in major repairs.

Buying a Used Machine

A secondhand tablesaw may be the ticket if you are looking for a Class A machine at a Class C price. This option will challenge your scavenging talents and your mechanical abilities, but you'll learn a lot about tablesaws. And you can end up with a machine that you'll never outgrow. But that's the subject of another article.

ROBERT M. VAUGHAN is a contributing editor to *Fine Woodworking*. He repairs and restores woodworking machines in Roanoke, Virginia.

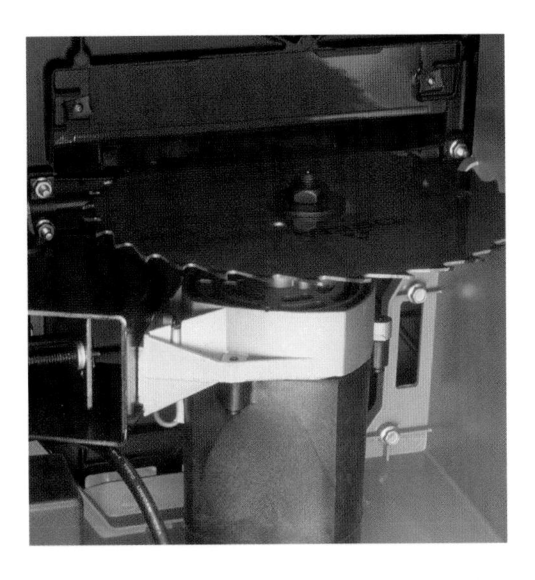

LIGHTWEIGHT TILT-AND-RAISE MECHANISMS are typical of low-end saws. Less powerful, direct-drive motors also are common in these Class E saws.

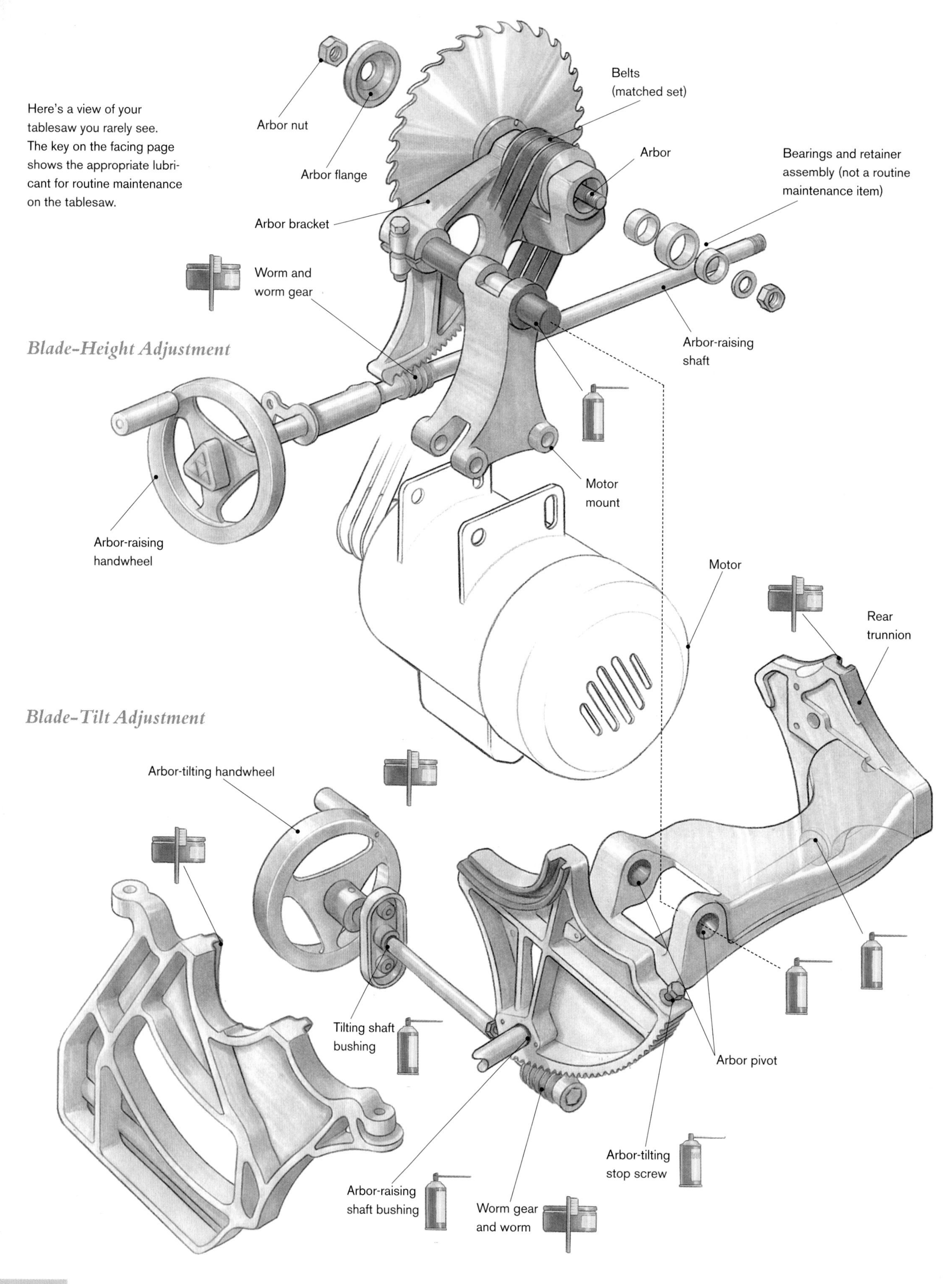
Here's a view of your tablesaw you rarely see. The key on the facing page shows the appropriate lubricant for routine maintenance on the tablesaw.
Arbor nut
Arbor flange
Belts (matched set)
Arbor
Bearings and retainer assembly (not a routine maintenance item)
Arbor bracket
Worm and worm gear
Blade-Height Adjustment
Arbor-raising shaft
Motor mount
Arbor-raising handwheel
Motor
Rear trunnion
Blade-Tilt Adjustment
Arbor-tilting handwheel
Tilting shaft bushing
Arbor pivot
Arbor-tilting stop screw
Arbor-raising shaft bushing
Worm gear and worm

Tablesaw Tune-Up

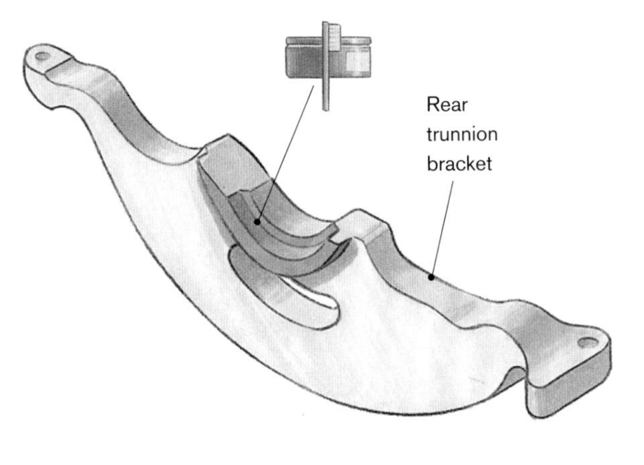

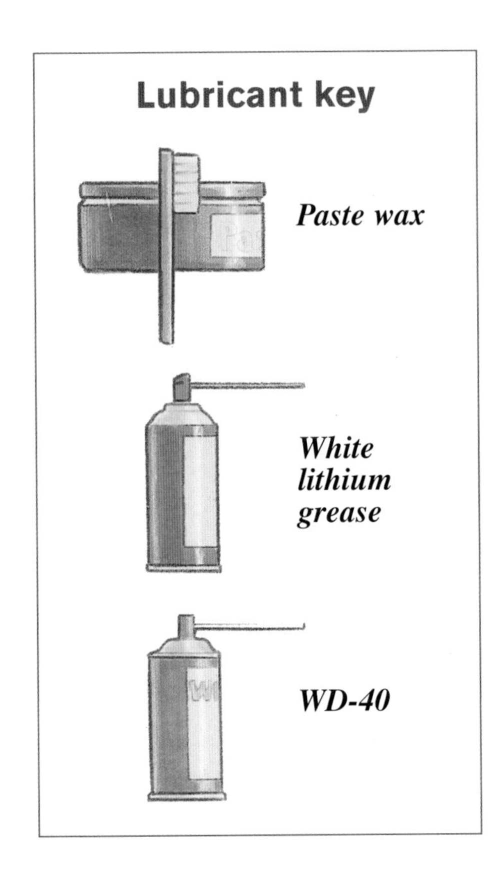

BY KELLY MEHLER

Several times a day, dozens of times in the course of a week, I crank the handwheels to adjust the blade on my tablesaw. Each time, the smooth, precise response from this otherwise ordinary task gives me a brief sense of satisfaction—things are okay. But as the months and board feet of wood slide by, the once silky-smooth operation starts to take more muscle. Eventually, tugging on the hand-wheel raises the blade in intermittent jerks, and tilting the blade provokes a metallic squeal. My saw is telling me it's time for a tune-up.

The tablesaw is the most important power tool in my shop. Accurate and heavy, it's built for the long haul. But it's easy for me to take it for granted. I routinely check the cutting accuracy, but I don't have a schedule for servicing internal parts. The cabinet base, valued for its stability, noise reduction, and dust containment, shrouds the motor and internals—out of sight, out of mind. So, even though I know that cleaning and lubrication keep the saw in top shape, it's only when I notice stiffness or noise while raising or tilting the arbor that I'm finally prodded into action.

The frequency of maintenance depends on how, and how often, the saw is used. Cutting abrasive materials, such as particle-

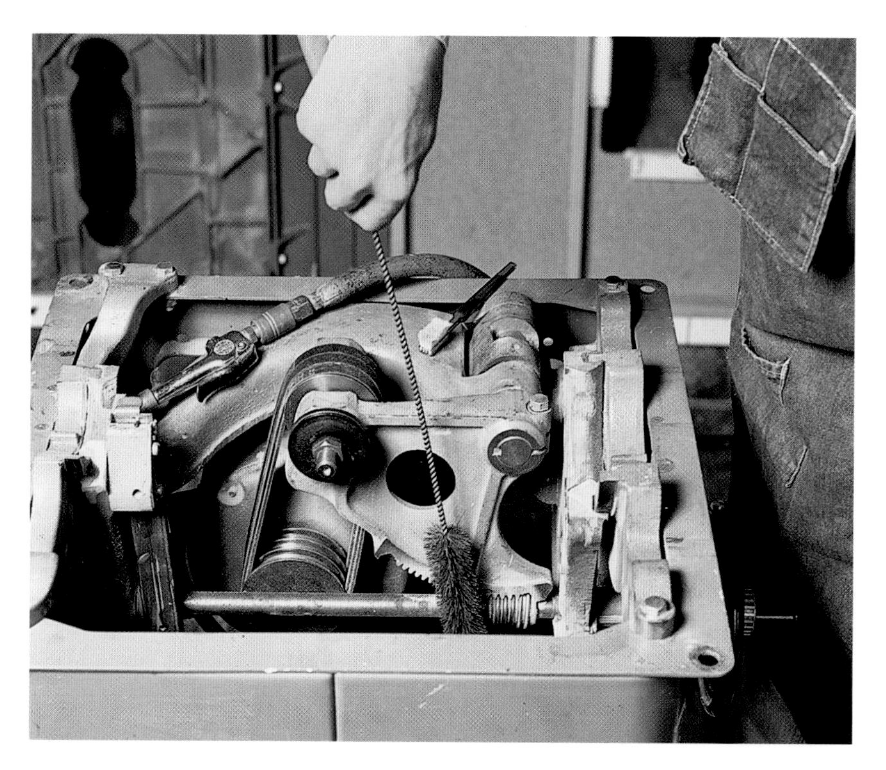

CLEAN OUT ALL THE ACCUMULATED SAWDUST **prior to inspection and lubrication.**

board, Masonite®, and Formica®, will increase the wear on internal parts. Sawing plenty of gummy, resin-rich or green wood creates pitch buildup. In my shop, internal parts should be cleaned and lubricated about once a year, and I set aside at least half a day to do it.

Tablesaw Anatomy

It's a lot easier to maintain your tablesaw if you have the original instruction manual and the parts list. All of the machine's parts usually are shown as they would be assembled, which can be especially helpful when doing repairs and replacing parts. If you don't have a manual and parts list, ask for one. Most manufacturers will oblige if you give the name and serial and model numbers of your saw. Manufacturers' addresses can be found in the Thomas Register® at your local library.

The drawing (on pp. 32–33) shows the guts of a typical cabinet-base tablesaw. The arbor assembly is the heart of the saw. It's a structural casting, with integral worm gear, that houses the sawblade drive shaft (the arbor) on a set of bearings. In addition, the motor, motor mount, belts, and pulleys also are part of this assembly. The trunnion assembly, also with an integral worm gear, supports the arbor assembly and allows the whole unit to tilt about the two arc-shaped slides, which are called trunnions. They engage mating brackets mounted to the front and rear of the cabinet. Handwheels control blade elevation and blade tilt by turning worms that engage worm gears on the arbor and trunnion assemblies. These parts work best when they're clean.

Remove the Top

Before doing any work on your saw, make sure it's unplugged. The best way to access all the internal workings is to remove the top from the saw. Removing a few screws at the upper corners of the cabinet is all it takes. But before you run off to get the wrench, you should measure and record the distance from the inside edge of the miter-gauge slot to the tip of the sawblade. Take this measurement with the arbor set at 0° and the blade elevated to its maximum height. This will aid you in getting the top back to its original position. If you've built jigs for your saw and they use the miter-gauge slot to reference their position relative to the blade, you'll want to replace the top exactly where it was.

Realigning the top can be a fussy, painstaking process. If you don't want to mess with it and the interior is not badly loaded with pitch, then most of the work can be done (with some difficulty) through the throat plate and the other openings in the cabinet. It's a personal preference.

If your saw is in dire need of a cleaning, remove the throat plate, the blade, the miter gauge, the fence, and any other loose items, and then remove the top. With the top out of the way, you can methodically work your way through the machine in a multistep process of cleaning, inspecting, and lubricating.

The Arbor Assembly

Cleaning the interior of your saw prevents the accumulation of pitch and sawdust, which increases wear and makes operation difficult. The first step is to clean out all the dust and gunk from all the moving parts. This will make inspection and lubrication easier (or possible). Use a stiff-bristle brush to knock loose sawdust from the arbor, arbor pivot, worm and worm gear. If your shop has an air compressor, a well-directed blast of compressed air really helps to clean hard-to-get-at areas.

Next you'll need to remove the accumulated pitch, gum, and packed sawdust. This is tenacious stuff, and you'll need some additional cleanup tools and solvent. A narrow putty knife, an old screwdriver, splints of wood, and a wire brush will help to dislodge the cakes of pitch and sawdust. The solvent I particularly like is Oxisolv® blade and bit cleaner (see "Sources" on p. 38) because it's nontoxic, nonflammable, and water soluble. It is as effective as oven cleaner without the noxious fumes, and it can be wiped off with a dry rag—no water needed.

The Arbor and Bearings The arbor needs very little maintenance, but you should check for burrs on the face of the arbor flange, which will cause the sawblade to wobble. Also, check the arbor threads for burrs and caked sawdust. A wire brush will remove the crud from the threads, and a fine-cut file can be used to remove the burrs.

Any wear or looseness in the arbor bearings also will result in sawblade wobble. To check the bearings, loosen the motor mount, and take the tension off the belts. Turn the arbor by hand, feeling for roughness. Grasp the arbor and gently pull up and down to check for any slack in the bearings. Temporarily remount the blade, and see if it spins freely. Roughness or slack in the bearings or failure of the blade to coast smoothly means the bearings need to be replaced.

Replacement is not routine maintenance. This involves removing the trunnion assembly, unseating the bearings, and replacing them using an arbor press—something probably best done at a machine shop or by a repair technician.

Blade wobble also can occur when the arbor flange is not perpendicular to the arbor. You can determine this by measuring the out-of-plane motion of the flange—this value is called runout. To determine the runout, use a dial indicator with a magnetic base. Mount the magnetic base to the closest rigid structure (the arbor bracket or the top if it's in place), and place the indicator tip against the flange. Rotate the arbor. Runout should be less than 0.010 in. More than that will cause enough vibration at the

PASTE WAX APPLIED WITH A TOOTHBRUSH **lubricates gears and doesn't attract sawdust.**

Tugging on the handwheel raises the blade in intermittent jerks, and tilting the blade provokes a metallic squeal. My saw is telling me it's time for a tune-up.

edge of the sawblade to cause rough cutting as well as splintering (especially with sheet stock). If the flange needs truing, remove the arbor assembly, and take it to a machine shop.

The Motor The motor runs in a dust storm inside the cabinet. Because of this environment, a quality saw has a totally enclosed fan-cooled motor (the motor windings and bearings are sealed within a steel shell, and an external fan blows cooling air over the motor housing). For long motor life, make sure this fan is free of obstructions, such as caked sawdust, on the intake grill.

To promote free air circulation, the cabinet has openings. Keep the level of sawdust in the cabinet to a minimum, well below the motor. If you have a motor cover on the cabinet, then the vents in the base should be clear of accumulated sawdust. Too much sawdust and pitch inside the saw base also is a fire hazard—another reason to practice good housekeeping.

V-Belts and Pulleys Most cabinet-base tablesaw arbors are driven from a motor via two or three V-belts, which are sold and installed as a matched set. Check for frayed or cracked belts, and replace them with new ones to the manufacturer's specifications. If only one belt is worn, replace them all as a set; otherwise, more of the load will be carried by the new belt. Uneven loading results in premature wear and vibration in the saw. Vibration transmitted to the blade causes rough cutting.

Pulley Alignment and Belt Tension The arbor and motor shafts should be parallel to each other, and the pulleys must be in alignment (see the drawings on the facing page). Even a slight misalignment will cause excessive belt wear from poor tracking and will increase vibration and noise.

To make this alignment, loosen the setscrew in the pulley on the motor shaft. Place a straightedge on the arbor pulley so that it makes contact with both edges of the rim, and then move the motor pulley until the straightedge touches both sides of its rim, too.

If the pulleys are aligned, then the shafts will be parallel. If you can't get the pulleys to align, it's because the shafts aren't parallel. In that case, loosen the motor mounts, and shift the motor until you get the desired alignment.

Once the pulleys are aligned, slide the motor mount to tension the belts. When you can deflect the belts about 1 in. at the center span between the pulleys using light finger pressure, the tension is correct.

Arbor Worm Gears and Arbor Pivot The arbor-raising worm and worm gear also are exposed to a blast of sawdust. They get packed with pitch and caked sawdust.

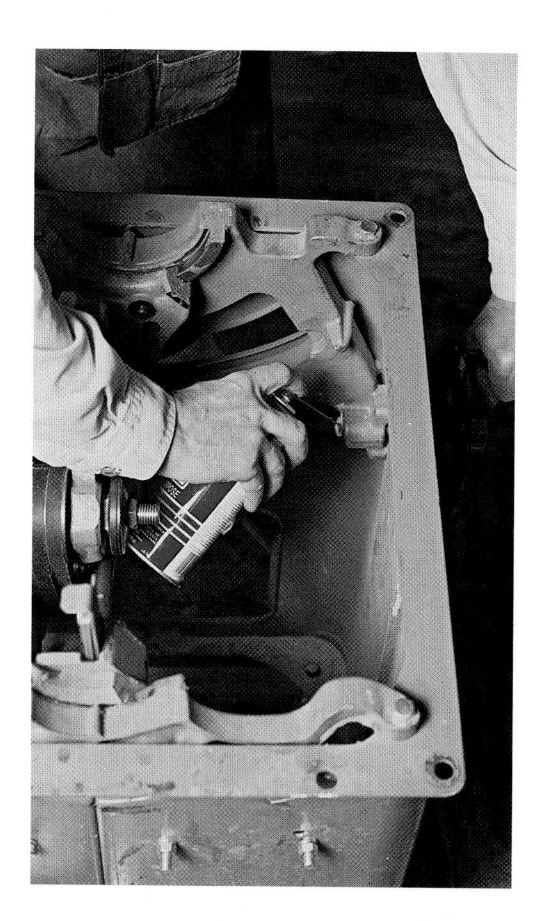

WHITE LITHIUM SPRAY GREASE is used to lubricate the hard-to-reach pivots.

A Tune-Up for Lasting Reliability

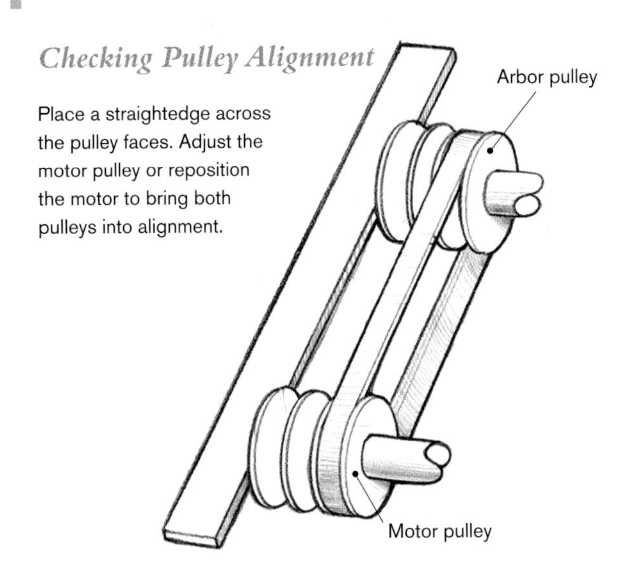

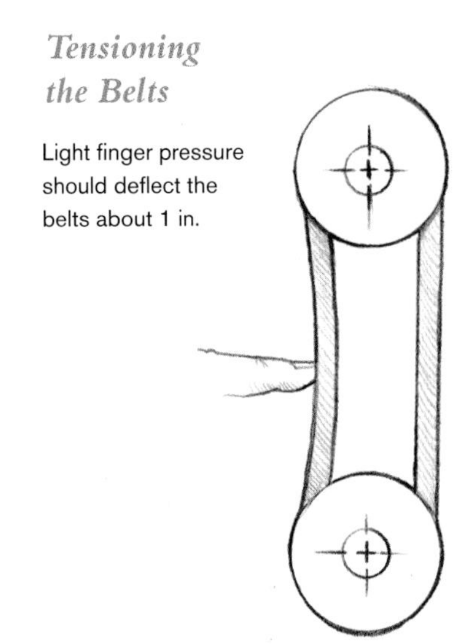

Enough of this stuff can make operation difficult. Use a stiff-bristle or wire brush to dislodge the material. For really tough cases, like pitch buildup, use Oxisolv cleaner.

The best lubricant is one that does not attract sawdust, such as powdered graphite, hard wax or white lithium grease. For the worm and worm gear, I use furniture paste wax. Use a toothbrush to work it into all the gear teeth (see the photo on p. 35).

Clean the accumulated gunk from the bushings that support the worm shafts. Strips of solvent-soaked rags used in a shoe-shine fashion work best here. Use this same technique for the arbor pivot. To lubricate these hard-to-reach areas, I use a white lithium grease spray (see the photo on the facing page). Then raise and lower the arbor several times to make sure that the operation feels smooth.

The Trunnion Assembly

Because the trunnions carry the weight of the entire arbor assembly, including the motor, they work best when clean and lubricated. Using your arsenal of cleaning implements, pick and scour the debris from the arc-shaped trunnion grooves and their mating trunnion brackets. Because you'll have to tilt the assembly back and forth to get it all, clean the worm and worm gear at this time, too. Using paste wax and a toothbrush, tilt the arbor assembly from stop to stop to work in the lubricant.

Arbor-Tilting Stop Screws The final step in the tune-up is lubricating the arbor-tilting stop screws. These usually are a hex-head machine screw with a locknut and need only a shot of penetrating oil, like WD-40® or Liquid Wrench®, on the threads. This will help them move easily when you set the arbor tilt for 0° and 45°.

Replace the Top and then Align

After the parts inside the cabinet have been cleaned and lubricated, put the top back in place. Reinstall the screws holding on the top until they are finger tight. Raise the arbor to its maximum height, and replace the sawblade. Because the slot-to-blade reference measurement was made with the blade set at 0°, you need to reset the blade to perpendicular, and set the stops. Then you can align the miter-gauge slots so that they are parallel with the blade.

Sources

Oxisolv Inc.
12055 Universal Dr.
Taylor, MI 48180
313-946-4440

Using a combination square or a draftsman's 45° triangle, set the arbor-tilting stops. Hold the square against the blade, and nudge the handwheel until the blade is perpendicular with the table. Carefully turn the machine screw until it's hard against the fixed stop. While holding the screw with a wrench, tighten the locknut. Tilt the arbor away, and then bring it back against the stop. Check again that the blade is still perpendicular to the top. Now, using the 45° part of the combination square or the triangle, tilt the arbor over, and repeat this procedure for the 45° stop.

With the blade at its maximum height, nudge the top back into its original position using your recorded measurement of the sawblade to miter-gauge slot distance. Snug up the cap screws using light torque on the wrench.

The Final Alignment Check This involves making a test cut. Clamp a ¾-in.- to 1-in.-sq. piece of hardwood to the miter gauge so that it extends about 1 in. past the sawblade. Cut the stock, turn off the power, and unplug the machine.

With the end of the workpiece at either the front or back of the blade, rotate the blade (backward so that no wood is removed) until you find the tooth that hits the wood the hardest and makes a scratching sound. Mark that tooth with a piece of chalk, and move the stock to the other side of the blade.

Look and listen as you rotate the blade. If the blade is parallel to the miter-gauge slot, the marked tooth will hit the wood and make the same sound. If the tooth does not hit the stock (it probably won't), the blade and miter slot are not parallel.

To jigger the top into the correct position, slightly loosen the cap screws holding the top to the base. Now tap the tabletop in the desired direction, as shown in the photo at left, rotate the arbor, and listen to the sounds the sawblade makes against the test piece. When the sounds match at the front and back of the blade, tighten the bolts and recheck.

Final tightening is best done in several go-rounds. If you crank down hard on the cap screws and go for the maximum torque in one yank, the phenomenon known as creep can throw the top out of alignment. So tighten the screws in steps, going from one to another, just as you would tighten the lug nuts when changing a tire on your car.

KELLY MEHLER is the author of *The Tablesaw Book* (Taunton, 2003) and makes furniture in Berea, Kentucky.

Shopmade Outfeed Table

ROLLER-TOPPED DRAWERS INCREASE OUTFEED TABLE CAPACITY. By extending the bottoms of two drawers at the back of his tablesaw, Frank Vucolo created a place to mount outfeed rollers. Here, he opens one drawer to rip a piece of 6/4 mahogany.

DRAWER SLIDE ALIGNMENT IS IMPORTANT. With the outfeed table flipped, the author positions a slide before he screws it to the poplar rail. Precise alignment ensures smooth operation of the outfeed rollers. A leg socket is below the square.

BY FRANK A. VUCOLO

In my small shop, ideal concepts are often compromised by the reality of limited space. My design for an outfeed table is a classic case in point. I started out thinking big. Ideally, I wanted the outfeed surface to extend 48 in. from the back of my tablesaw, so I would no longer have to set up and then reposition unstable roller stands. My ideal was quickly squashed, however, when I realized I couldn't dedicate that much permanent floor space. I need the space behind the saw to store my planer and router table when I'm not using them.

After some careful measuring, taking into consideration where I would locate all the machines, I concluded that the outfeed table should extend 30 in. from the back of the saw. But I still needed more support to rip long stock and to cut sheet goods.

While I was pondering possible solutions, I started to think about rollers that could extend off the back of the fixed table and then retract into it when they weren't needed. Then I remembered how amazed I was at the strength of Accuride's® extension drawer slides (150-lb. capacity) when I had used them for file drawers in a desk pedestal. After a little more head scratching, nudged along by a couple of cups of coffee, I decided to incorporate the slides into a pair of drawers with rollers mounted on the front of them for the outfeed table (see the photo at left above). Now I simply open a drawer to get an additional 24 in. of outfeed surface when I'm ripping long boards or cutting sheet stock.

Design and Materials

Allowing an extra 1 in. for the extension rollers and the drawer slide action, the outfeed table is designed to support work up

Outfeed Table Assembly

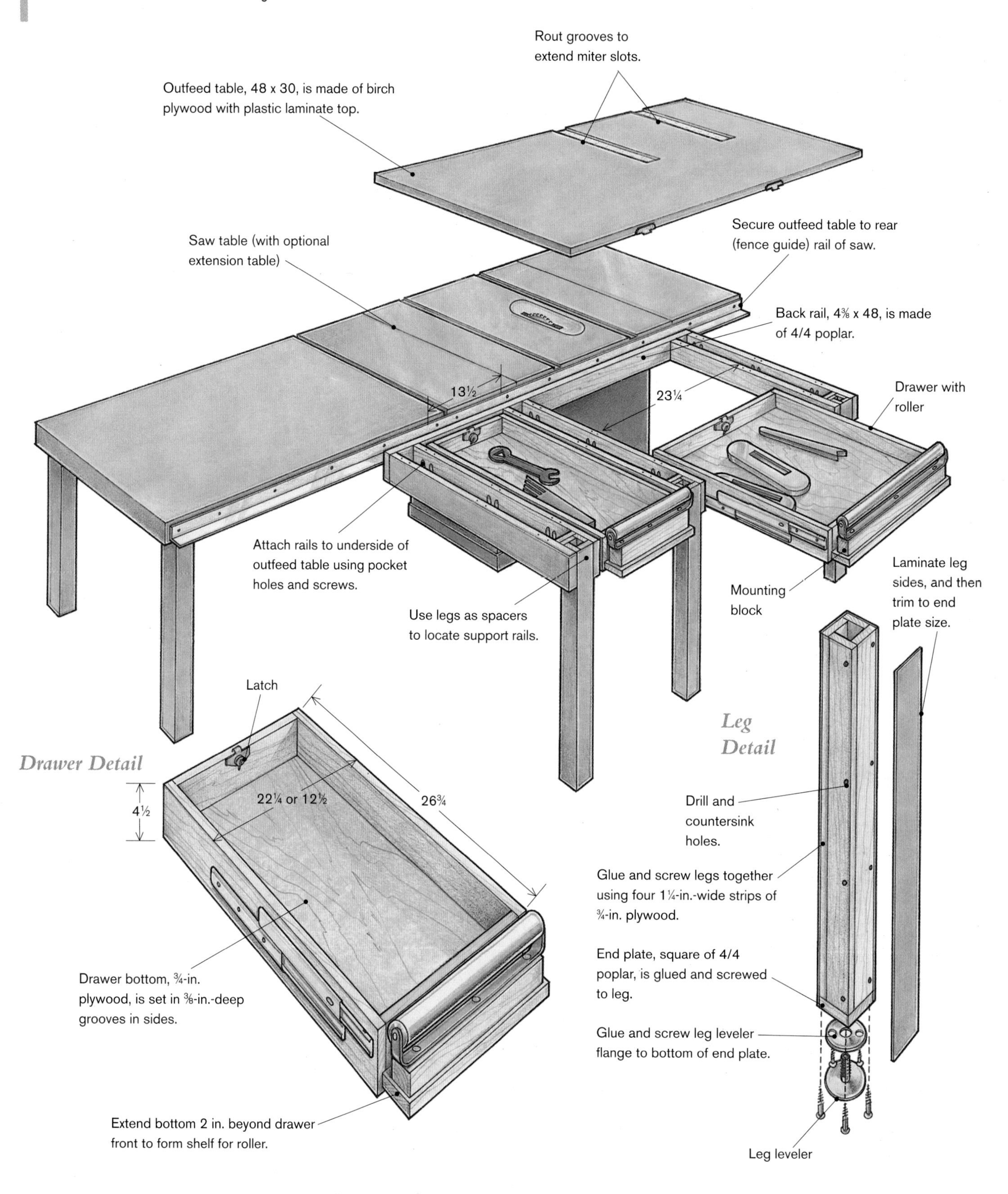

to 55 in. from the back of the saw table. With the drawers in the closed position, only 30 in. of floor space behind the tablesaw is committed. I made the drawers different widths so that I have various outfeed options, and I extended the drawer bottoms out in front of the drawers. This way, I have a place to mount the rollers (see the detail on the facing page). As a bonus, I get two drawers for storing saw accessories. And because the rollers are an integral part of the outfeed table, they are adjusted precisely in relation to the tabletop.

I constructed the outfeed table's top, legs, and drawer bottoms out of ¾-in. birch plywood. The under-table support rails are made from 4/4 poplar, as are the drawer sides, fronts, and backs. For added protection and to give a nice slick surface, I covered the legs and top with plastic laminate.

To complete the material requirements, I bought the following hardware: two metal rollers, one 13 in. long and one 22 in. long, two sets of heavy-duty drawer slides (I picked up Accuride's file-cabinet model from The Woodworkers' Store℠), three leg levelers (available from Woodworker's Supply Inc.℠) and a couple of latches (window sash locks), which I bought at a local hardware store (see "Sources" on p. 42). When you're determining the size of your drawers, keep in mind that the slides come in 2-in. increments, 12 in. to 28 in. long.

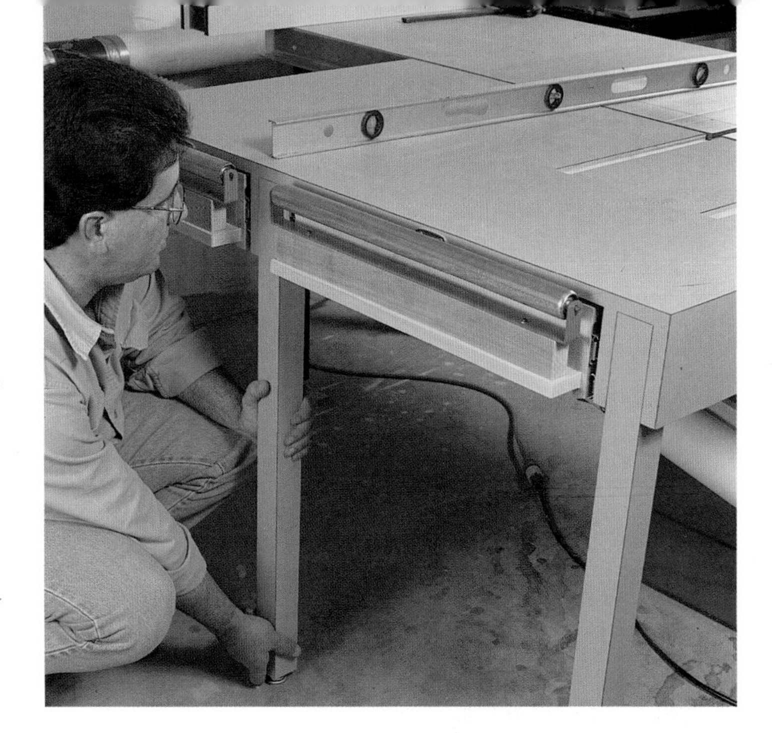

LEVEL THE OUTFEED TABLE TO MATCH THE SAW TABLE After Vucolo secured the outfeed table to the rear guide rail of his saw, he turns the leg levelers (screw feet) to line up the two surfaces.

Making and Mounting the Table

To build the outfeed table, first determine the overall size (mine is 48 x 30), and then cut the tabletop out of plywood. Temporarily mount the plywood to your saw, and level it using braces. This is so you can determine the length of the three legs. Measure each leg separately, and allow some room (½ in. or so) for height adjustment. The leg levelers will take up the play. Disassemble the table, and then fabricate the legs, as shown in the drawing detail on the facing page, including the plastic laminate.

With all three legs complete, lay out the support rail locations on the underside of the plywood top. Approximate the two different widths of the drawers plus their slides. Rip and crosscut the poplar pieces to size, and begin fixing the members to the plywood. I drilled pocket holes and then glued and screwed the rails in place. Start at one end, then use an assembled leg as a spacer to set the second rail. Next do the other end of the table, using another leg as a spacer. Set the two center rails in a similar fashion. Then attach the rear rail across the ends of the support rails. Also, cut and attach blocks behind each leg using the leg as a guide.

Mount the carcase portion of each drawer slide to the rails (see the photo at right on p. 39). Make sure you position all the slides the same distance from the bottom of the table. I used the rails as a reference. The drawers must be perfectly parallel to the top. While you have the table flipped, laminate the sides of the top, and trim them with a flush-trimming bit in a

POCKET HOLES AND SCREWS JOIN DRAWER BOXES. After temporarily clamping a drawer back, the author drives three screws into the sides using a flexible-shaft extension for his drill.

Sources

Wilke Machinery Co.
3230 Susquehanna Tr.
York, PA 17402
800-235-2100
metal rollers

The Woodworkers' Store
21801 Industrial Blvd.
Rogers, MN 55374
800-279-4441
drawer slides

Woodworker's Supply Inc.
1108 N. Glenn Rd.
Casper, WY 82601
800-645-9292
leg levelers

router. Turn the table over, so you can laminate and flush-trim the top.

Now temporarily mount the legs, and align the laminated table to your saw exactly as it will be positioned in use. Carefully mark the position of the miter slots on the top.

Determine the depth of the grooves by referencing off the tablesaw. If you have a T-slot or dovetail-shaped miter-gauge runner, lay out the slots so that they will be a bit wider than the widest part of the tablesaw slot. The outfeed table slots will be for clearance only.

Remove the outfeed table. Run the miter gauge all the way past the blade, so you can find the length of the runner as it hangs off the back of the saw table. Mark this length plus a bit extra onto the outfeed tabletop. If you use sliding jigs, like a crosscut box, check that their runners will work in the laid-out slot, too.

Using a straight bit and your router, cut the grooves in the surface of the outfeed table. A straightedge can be used to guide the router. But don't try to cut the whole depth in one pass. It's better to make two or three passes, removing a little at a time. Soften all the corners of the laminated top using a fine file. Also, ease the edges of the miter-gauge slots, and feather the edge that will go against the tablesaw. This will ensure that workpieces won't get hung up as they slide from the tablesaw onto the outfeed table.

How you mount the outfeed table to the saw will depend on the type of saw and fence guide rail you have. You can use angle brackets or drill directly into the rail. After you have the outfeed table in its approximate position, use a straightedge and a level to adjust the screw feet until the outfeed table is lined up to the saw table (see the photo on p. 41).

Adding the Drawers and Extension Rollers The drawers should have a ¾-in. plywood drawer bottom extending 2 in. beyond the front of the drawer. This will provide enough rigidity for the extension rollers (see the drawing detail on p. 40). To receive the bottom, I plowed a ⅜-in.-deep groove down the inside of each drawer side using a dado blade in my tablesaw. After I glued and screwed the bottom to each drawer, I butt-joined the front and back pieces together using pocket holes and screws (see the photo at left). Then I attached the other part of the drawer slides to the outsides of the drawers.

It's critical that the rollers are mounted at the correct height. They should be at, or just barely above, the outfeed surface; they need to roll freely, without disrupting the travel of a workpiece. To get the proper height, I mounted the rollers using spacer blocks. First I set the roller on the shelf created by the extended drawer bottom. Then I measured from the top of the roller to the tabletop. I cut the block a bit oversized and then planed it down to thickness. If the roller is not parallel to the outfeed top and you can't adjust the drawer slides enough, taper the blocks slightly with the plane until the top of the rollers are level with the table. Finally, install a latch on the inside back of each drawer, so you can lock them in the open position.

FRANK VUCOLO builds furniture for his home in East Amwell, New Jersey.

Jointer Savvy

BY BERNIE MAAS

Like good hand-tool usage, successful wood machining involves working by eye, feel, and intuition. Though fences and motor-driven cutters help, they can't do it all. A squeeze here and a push there—some call it "body English"—can make all the difference in truing an edge or flattening a surface. These hard-to-describe nuances really boil down to having savvy for the tool you're using.

The jointer is a perfect example of a savvy-demanding machine. To joint cleanly and consistently requires real proficiency. I've run a jointer for 35 years, and while it looks like a set-it-and-forget-it monster, it's not. So to save you from its pitfalls, I'll pass along the jointer pointers I've picked up and now share in my classroom. In addition to describing a jointer's makeup and telling you how I straighten an edge, I'll describe how I read a board's shape and grain, so I joint squarely without tearing wood—all while keeping my fingers safe (see the photo at right).

Jointing 101

Before you begin surfacing a pile of lumber for a project, it helps to be familiar with a jointer's components and how to adjust them. For reference, the drawing on p. 44 shows the machine's basic parts, their purposes, and the fundamentals of feeding stock, which are important during both simple and advanced operations (like the ones in the sidebar on pp. 48-49).

A jointer's main jobs are flattening faces of boards and squaring and straightening edges. These tasks are achieved by adjusting the jointer tables and fence in relationship to the cutterhead. If the fence is set at 90° and the outfeed table is set parallel and exactly to the maximum height of the

GOOD JOINTING REQUIRES LOOK AND FEEL. Before Bernie Maas edge-joints a board, he sets the jointer for a light cut, and then he reads the stock's shape and grain direction. While positioning the work to correct for any crook, he orients the grain to reduce tearout. Applying easy pressure with his fingers above a properly working guard contributes to the machine's safe operation.

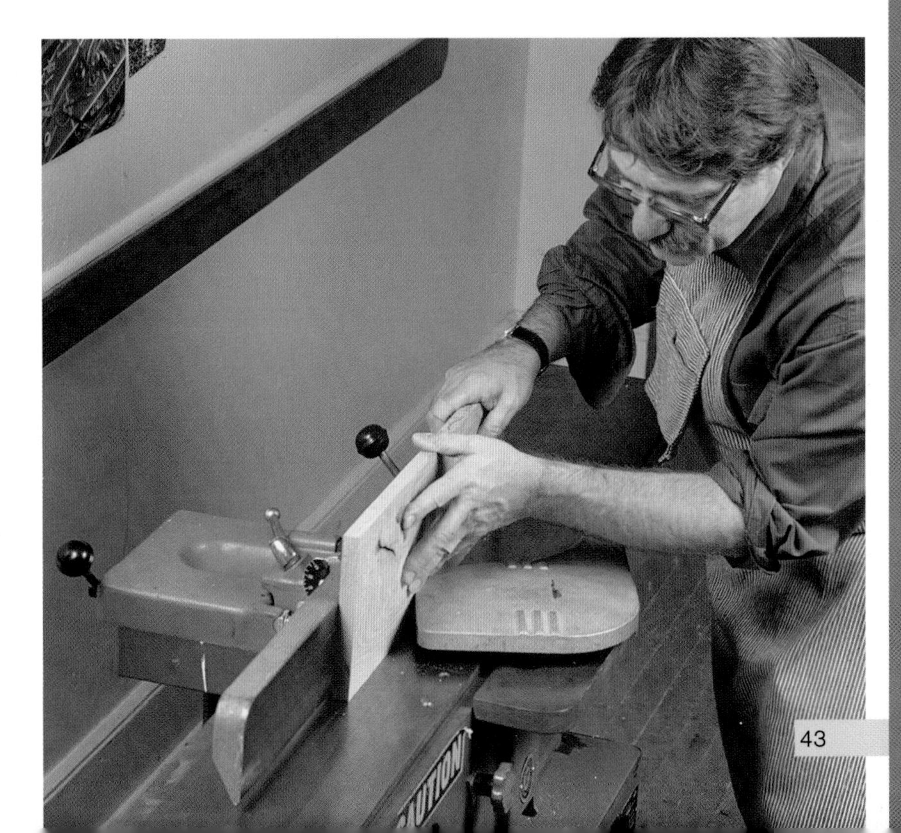

knives, you can establish a uniform cut by lowering the infeed table by the depth of cut you want. After you have a handle on jointer fundamentals, the most important item is safety.

Safety Comes First There are a few keys to jointing safely. First, be alert and focused. In addition to watching your work and your fingers, make sure your jointer's guard is in place and working properly. Second, work in a warm shop. Cold hands mean numb fingers that are slow and don't grab well. Third, use a push block when surfacing the face of a board. Fourth, don't joint a board less than 12 in. long or less than 1 in. thick. Finally, don't "white-knuckle" it. Feed with just enough force to advance the stock. This way, should the work kick or shatter, your hand won't go flying into the knives. If your fingernails are blue, you're pushing too hard. Either your bite is too big, your stance is off, the blades are dull, or your machine is out of whack.

Reading a Board's Grain Ideally, there will be only one grain direction per face of a board. If this is the case, you can easily prevent tearout while jointing. Just feed the stock so that the grain lines run in the direction of the cutterhead rotation—from high to low (see the drawing below). Of course, life isn't always that pretty: There are knots, and often the grain changes its mind. Sometimes the fibers run in three directions at once (figured cherry is a prime example of this). And it's more likely that tearout will occur the steeper the angle of grain reversal.

Jointer Anatomy and Feeding Stock

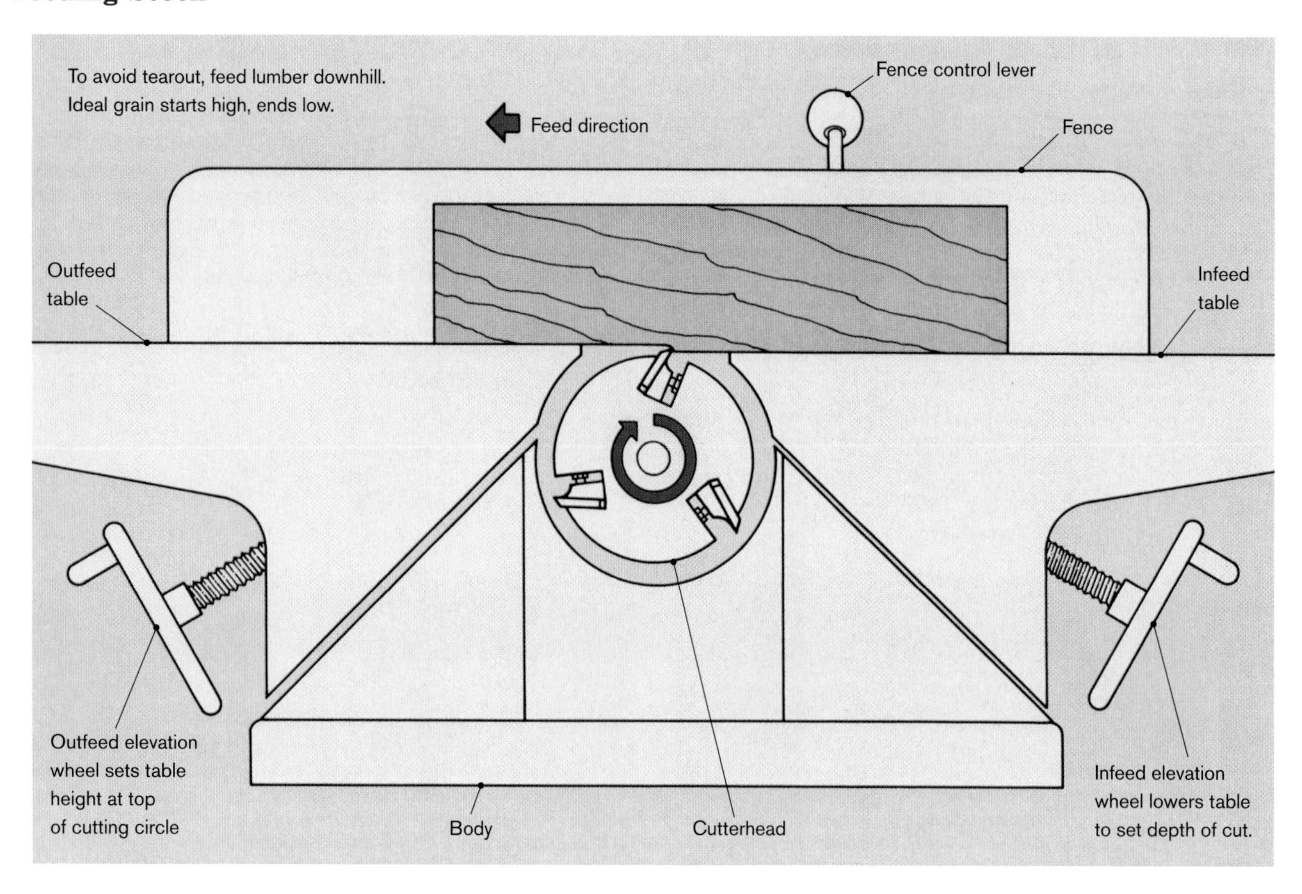

IMPROPERLY ALIGNED WORK. A beveled edge produces a gap between the fence and the stock's face. In this case, the infeed table guides the workpiece, so jointing will only reproduce the same bevel instead of squaring the edge of the board.

CORRECTLY ALIGNED WORK. Proper pressure and a bit of "body English" snug this board squarely to the fence. The wedge-shaped gap above the infeed table may point away from the fence, as shown, or toward it, depending on the grain direction.

So how do you handle recalcitrant stock on your jointer? First, make sure the knives are sharp and protrude equally from the cutterhead, or reserve a sweet spot—a section of the knives dedicated for when you need a razor edge. (I keep a sweet spot in a rarely used space at the rear of the machine.) Second, remember that a fast cutter means a clean cut. Think of your jointer as a lawn mower. When you push the mower into tall grass, the blade lugs down and rips the grass instead of shearing it. To solve this problem, you raise the blade to take a shallower cut, and you go slowly so the engine stays revved. You can joint stock with nasty grain the same way. Reduce the cut to $\frac{1}{64}$ in., and then move the problem area of wood slowly over the knives. Keep the knives whirling at top rpms by slowing the feed rate. Although you'll need more than one pass, the cut will be smooth. If you're still having tearout trouble, you may want to back-bevel the knives (see the sidebar on pp. 48–49).

Squaring Stock When jointing, you should always square an edge to an established face. To do this, joint the face of the board. Then using the newly jointed face as the reference, joint one edge square to it. If your edge is still out of square and you've double-checked your fence setting, most likely the jointer table is guiding the unsquare edge of your work. By guiding the work this way, you're not correcting anything; you're only duplicating the existing angle. To avoid this problem, hold the piece tightly against the proper reference—the fence. There should be a tiny wedge-shaped gap between your work and the infeed table. Now feed the work over the knives. The edge should be square.

Correcting a Board's Shape Unless you enjoy having lumber bind up in the tablesaw, you will want to remove any crook in a board before you rip it. (Crook is a curvature that runs from end to end along the edge of a board.) Although crook is common, getting rid of it isn't hard. If your project doesn't require the full length of a board, crosscut it first. Shorter pieces are easier to handle, and they produce less jointing waste. Try to joint with the concave (hollow) edge down. The work should be stable because it'll rest on the "ears" (ends of the arc) as it passes over the knives.

Sometimes, however, you'll have to joint a piece with the convex edge down. Because it will want to rock, you'll be tempted to hold down the piece tightly to the infeed table as you feed it through the jointer. But if you do this, the board will come out as crooked as when it went in. Instead, set the jointer for a ⅓₂-in. cut, and balance the board on a high spot. (The different heights of the two tables should help steady the piece.) Apply moderate pressure, and feed through the knives to create a flat on the edge. Repeat until the work becomes stable as it seats onto the tables. Gradually, you'll even the flat along the entire length.

STRAIGHTENING THE EDGE OF A PANEL. **To square and straighten panel edges, the author draws an ink line and then shaves toward it with his jointer. After he aligns the work by eye, so the line is parallel to the table, he takes a couple of passes to create a flat. By continuing, he will true up the entire edge.**

JOINTING END GRAIN. **To avoid splitting the end of a stile, Maas used his belt sander to chamfer the corner before jointing. By carefully easing the frame over the cutterhead, he will flatten each of the four edges. Note that the grain in this rail is running slightly downhill, in the direction of the knife cut.**

Straightening Edges and Panels You can modify convex-jointing principles to straighten a board's edge parallel to its opposite side; in which case, you'll be wasting a small wedge shape. Or you can shave a bit off a panel to make it fit a tight frame. This is great for getting a plywood back to fit an out-of-square carcase that diagonal pipe clamps won't yank back. Start by marking the cut line. I use a fine ball-point pen, which leaves a crisp, dark line that shows up well under shop lights. Make sure you mark the line on the side of the board that will face you as you feed (see the top photo at left). Set the jointer for a light cut (1/64 in. to 1/32 in.), and take a pass. If the taper amount is slight, apply less pressure on the thin end. If you're beginning the cut at the thin end, start with easy pressure, and increase the squeeze as you go. If there's a considerable amount to taper, lift the thin end until you can eyeball the line parallel to the infeed table. Lightly joint a flat parallel to the taper line. To get the flat back on target if you've strayed, take partial cuts. Your final pass should run the length of the panel for a clean, smooth edge.

Smoothing Tapers Lots of folks have jigs for ripping tapers on the tablesaw. I've done loads of them that way, but I've always felt that it was a kickback waiting to happen. Instead, I prefer to bandsaw and joint my tapers. I start with rough stock that's about 1/4 in. wider than the maximum taper I need. The extra width gives latitude for truing up later. I mark the stock with two lines. The first is the exact taper line. (For safety, I never taper to less than 1 in.) The second line starts ¼ in. below the first. Next I bandsaw as close to the second line as I can. Then I set my jointer for a light cut. Feeding the thick end first with the grain running downhill, I pass the piece over the cutters until I reach my taper line. The best part is that the jointer won't leave ridges and burns as sawblades will. Though it is somewhat tricky, a few furnituremakers cut tapers using the jointer alone (see the photo at left on p. 48).

Jointing End Grain With care, you can use the jointer to straighten the end grain edge of a panel. If the panel's trailing edge is unsupported, though, it's likely that a ½ in. or so of end grain, which has no structural integrity, will separate from the stock (with a loud whack) as soon as the knives hammer away at it. Luckily, there are several ways to deal with this problem. You can clamp a back-up piece to support the trailing edge before jointing. You can joint half one way and then half from the opposite end. Or you can nick the exiting-end fibers with a utility knife just above the height of the cut. However, my favorite solution came from my eighth-grade shop teacher who simply chamfered his trailing edges slightly above the cut. You can use a rasp, a block plane, or a belt sander to make the chamfer, and the method works equally well on frames, as shown in the bottom photo on the facing page. Any chamfer that's left after jointing will feather away when you finish-sand.

Rabbets, Bevels, Chamfers Chances are good that your jointer has a rabbeting arm, which will support a workpiece at the end of the jointer knives. A ledge in the outfeed table allows you to plow rabbets because the rest of the board will pass by the cutterhead unimpeded. The rabbeting feature of a jointer is appealing when you consider the setup time required to do the job on a tablesaw or router table, but be mindful that a jointer's ledge size determines the largest rabbet you can make.

CLEAN RABBETS. **With the fence set near the jointer's edge, the author slides a piece along the rabbeting arm, taking 1/16 in. at a pass. For clean rabbets, all the knives should project equally toward the rabbeting-ledge end of the cutterhead. A few woodworkers sharpen the ends of the knives to further improve the cut.**

SMOOTH CHAMFERS AND BEVELS. **A jointer is more versatile than you might think. For decorative table legs, for instance, you can chamfer the corners on the jointer. Or you can precisely bevel staves for a planter. Just angle the fence, adjust for a shallow cut and then feed the work edge over the jointer in several passes.**

Finally, a jointer-made bevel or chamfer adds a nice detail to almost any project. For example, to neatly dress the front of your bookcase shelves, set your jointer fence to 45°, adjust the infeed side for a 1/16-in. cut, and then pass each shelf along the fence as you go over the cutter (see the photo above). Without a square edge, the shelf is more pleasing to the eye and to the hand.

BERNIE MAAS is an associate professor of art at Edinboro University of Pennsylvania, where he teaches woodworking and computer-aided drafting. In his free time, he likes to build furniture for his home.

Using a Jointer: The Advanced Class

Back-Beveling Jointer Knives

Slightly back-beveled jointer knives can reduce tearout and leave a smoother surface on highly figures woods.

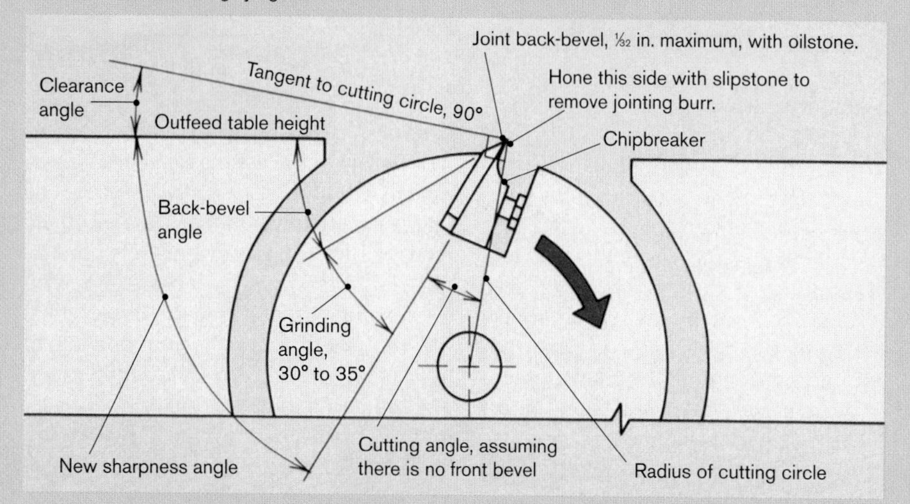

Assuming your knives are properly set, there's a refinement called back-beveling, which can improve the quality of the cut especially on figured woods. I'll describe measures to reduce tearout, including how I back-bevel knives. But first I'll explain what down pressure and jointer-table adjustments will do for you.

Down pressure and table settings: For straight jointing, first apply infeed pressure followed by pressure on the outfeed side, but for a tapered edge, keep pressure on the infeed table only (see the photo at left below). To taper a leg, for example, I mark the taper beginning 1 in. to 2 in. back from where I want it to start. (I straighten the small starting arc with a plane later.) With the leg butting a stop clamped to the infeed table, I take 1/16-in.-deep passes while maintaining infeed pressure with a push block. I count the number of passes so that I can match the taper on all sides of the leg.

To create a "sprung" joint (edges that are jointed with a 1/64 in. or less hollow in their centers to aid in clamping a glued-up panel), I slightly increase downward pressure at the middle of each edge as I pass it over the head. On certain jointers, you can also spring an edge by readjusting the outfeed table. To help me understand table alignments, I consult *Woodshop Tool Maintenance* by Beryl Cunningham and William Holtrop (the book is out of print, so check used-book websites). The drawing on the facing page, which is adapted from the book, shows what effects various table heights and pitches will have on a jointed surface.

Reducing jointer tearout: When a jointer is running, slight changes can occur in the cutting arc due to distortion in the head or bearing play. As a result, one knife may project farther than the others and tear the surface. The tearout will be most noticeable on figured woods. There are several ways to combat tearout, including skewing the work, proper sharpening, and back-beveling the knives.

Sometimes when I'm jointing wood with difficult grain or high figure, such as bird's-eye or curly maple, I use an auxiliary

TO TAPER STOCK, use pressure on the infeed table. When tapering a leg, clamp a stop on the infeed side to align the start of the cut; then advance the work over the knives (the guard has been removed for clarity).

AN AUXILIARY FENCE SKEWS the work to leave a smoother surface on some figured zebra wood. Tischler screwed the plywood fixture together, and then he shimmed and clamped it to his jointer's fence.

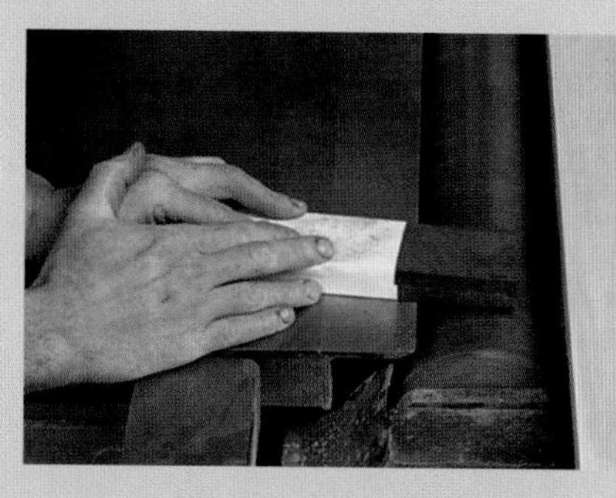

CAREFULLY BACK-BEVEL THE KNIVES. Clamp a stop block to the infeed table. Then wrap an oilstone with paper. Set the outfeed table, so the stone evenly grazes the knife tips once the jointer is running.

fence to skew the work (see the center photo on the facing page). Skewing, like angling a plane, can reduce tearout. An auxiliary fence works best on a wide jointer that allows more skew.

The primary bevel on a jointer knife (see the drawing on the facing page) can range from 30° for softwoods, to 35° for hardwoods. To get the angle, I send my knives to a grinding shop because they precisely grind a flat bevel as opposed to a hollow bevel. Occasionally, I'll add a secondary (front or back) bevel myself. This secondary bevel increases the sharpness angle (by adding the back-bevel angle to the grinding angle). But I will only do this if the knives are starting to dull or if I need to surface a run of difficult stock.

It's difficult to grind a front bevel. But back-beveling, as described in Charles Monnett Jr.'s book, *Knife Grinding and Woodworking Manual* (Charles G. Schmidt, 301 W. Grand Ave., Montvale, NJ 07645), can be done with the knives in the cutterhead. Jointing the back-bevels also removes any cutting-arc eccentricities at the same time (see the drawing on the facing page).

TO REMOVE THE BURR, use a slipstone lubricated with oil. With the jointer disconnected, hone the underside of each knife equally, being careful not to cut a finger as you guide a fine India stone.

Back-beveling is dangerous and should be done only with the following precautions: The knives must be accurately ground, balanced, and set. The outfeed table should be barely lowered—just until the knives whisper a ticking sound when an oilstone is passed over them as they're rotated by hand. Wrap the stone in paper, so you won't scratch your outfeed table. Finally, clamp a stop across the infeed side near the cutterhead (see the photo at right on the facing page) in case the stone kicks.

Once you've safely set up the job, lower the infeed table just below the outfeed table. Start the jointer, and allow it to reach full speed. With a firm (but not monkey- tight) grip, slide the oilstone across the outfeed table as you dress the knife tips. Turn the machine off, so you can examine the knives. You have ground a proper bevel if there's a thin, straight gleam on the edges. If you didn't get enough back-bevel, lower the outfeed table another 0.001 or 0.002 in., and repeat the jointing. Your back-bevel shouldn't exceed 1/32 in.; any more than this will cause the heel of this secondary bevel to pound into the wood.

After back-beveling, hone the knives to remove the burr raised by jointing. To do this, I use a fine India slipstone (see the photo at left). First I unplug the machine. Then I carefully work the oiled stone back and forth across the knife face that protrudes from the chipbreaker. I hone all the knives equally, stopping when I no longer feel drag with the triangular stone. Your patience with this procedure will be rewarded by the cut you'll obtain.

PETER TISCHLER is a North Bennet Street School graduate who runs a chairmaking and cabinetmaking shop in Caldwell, New Jersey.

Effects of Infeed and Outfeed Table Positions

1) Outfeed table high in center results in concave surface.

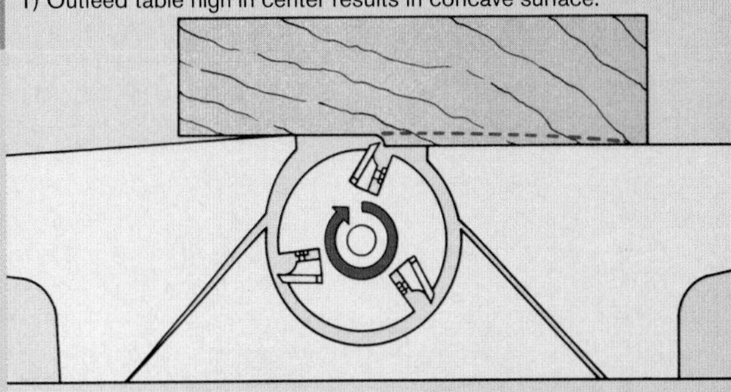

2) Low parallel outfeed table results in heavy cut at back end.

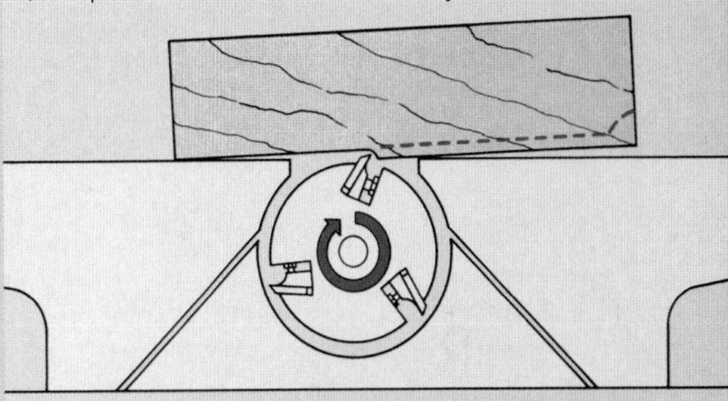

3) Both tables high in center and nonparallel result in concave surface.

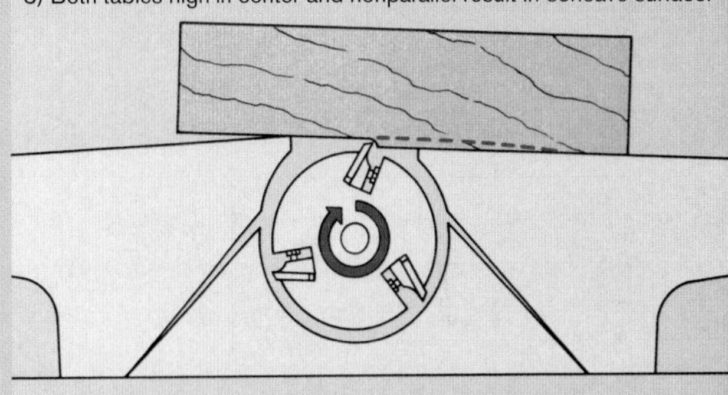

4) Both tables low in center and nonparallel result in convex surface.

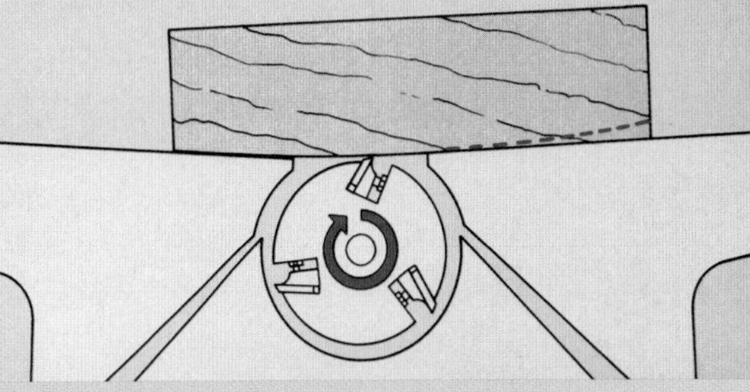

Jointer Tune-Up

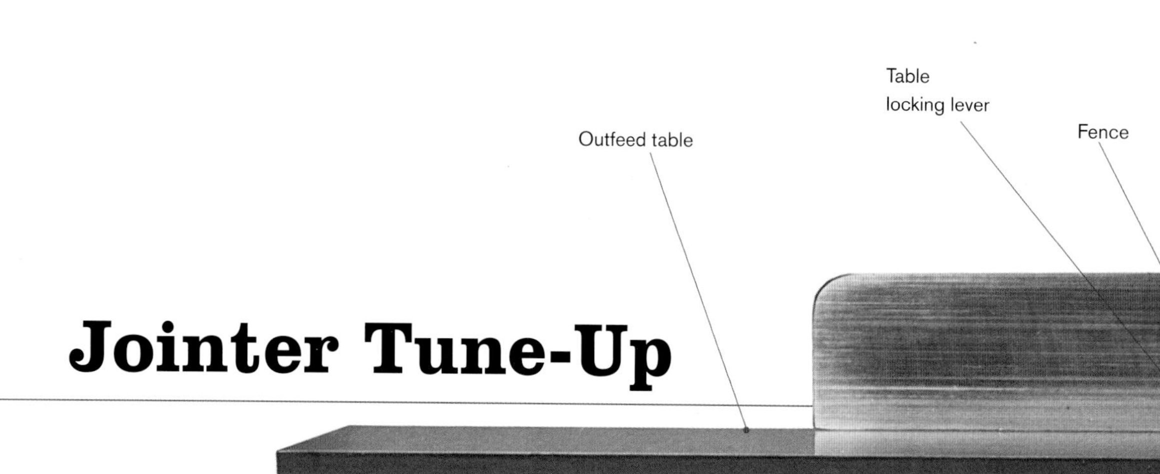

BY JOHN WHITE

Truing rough stock begins at the jointer. But an out-of-tune machine that snipes the ends of boards or mills curves into stock isn't of much use. Because of wear, damage, or imperfect castings, jointers may become misaligned over time. Fortunately, most machines can be adjusted without a lot of specialized equipment or mechanical skills. And while you're at it, consider replacing dull knives, a task many woodworkers attempt only in moments of desperation.

Jointers are relatively simple tools. The infeed and outfeed tables flank a cylindrical cutterhead containing three knives. The tables on most small and midsized jointers move along sloped dovetailed ways, which are wear surfaces. Over the years the tables may begin to droop. Occasionally, jointers fresh from the factory may exhibit these bad traits, too.

The infeed table and fence guide the stock as it crosses the cutterhead. The outfeed table picks up the freshly jointed surface and guides and supports the stock as the pass is completed. The jointed surface is only as straight as the path the wood takes across the cutterhead. If the tables slope, the wood follows the same path. If the tables are misaligned, stock may have a snipe (a deeper cut) or a hump (an uncut section) at the end of the cut or a curve along its length.

The basic tools required for a tune-up are a set of feeler gauges, a small try-square, and a good, short straightedge such as the blade of a combination square. A 6-in. dial caliper may come in handy for gauging shim stock, but the job can be done without one. To check the tables for flatness, you'll need a long machinist's straightedge or a test bar (for directions on making and using a test bar, see p. 55) to span the length of both tables.

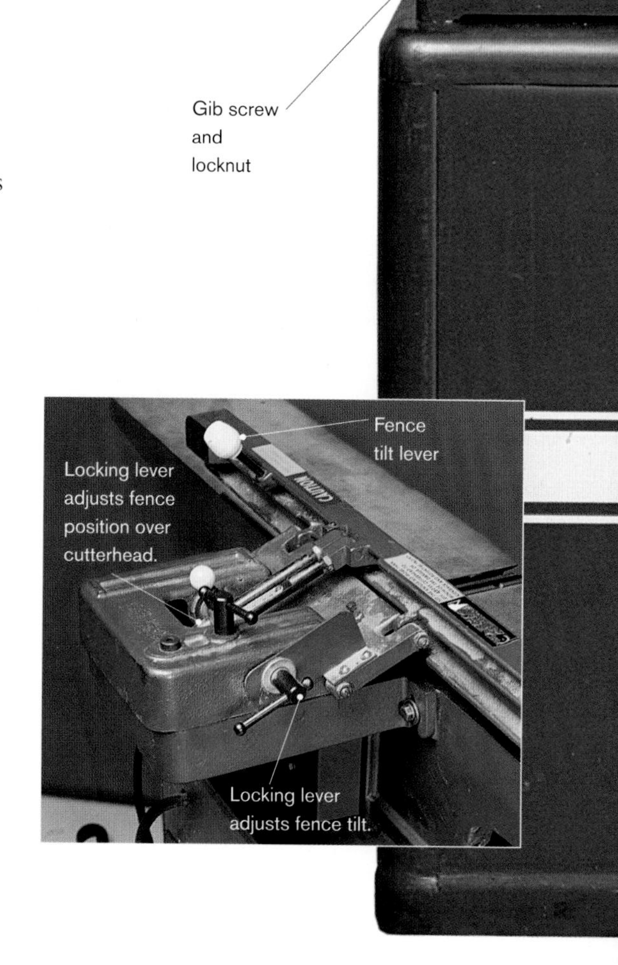

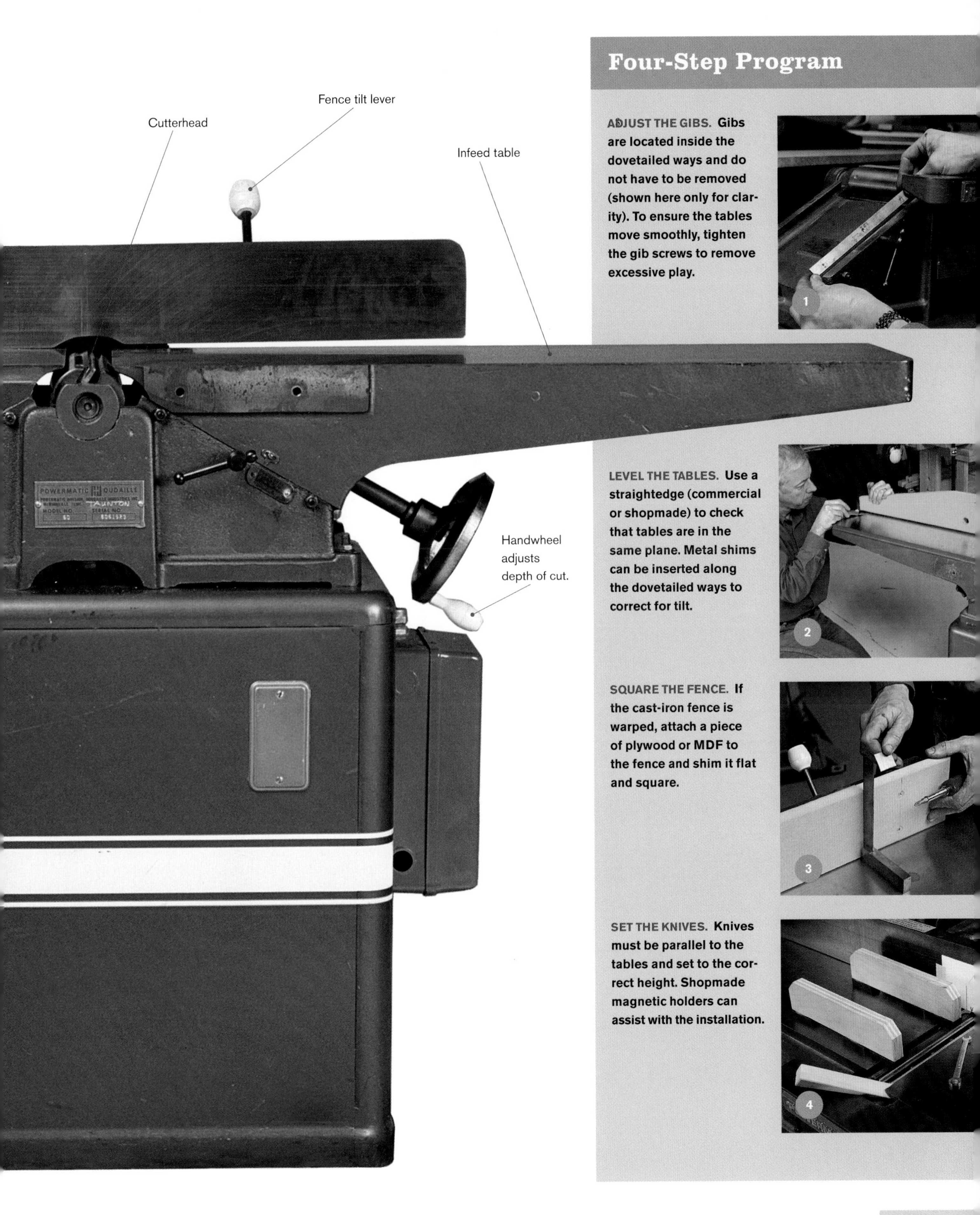

Four-Step Program

ADJUST THE GIBS. **Gibs are located inside the dovetailed ways and do not have to be removed (shown here only for clarity). To ensure the tables move smoothly, tighten the gib screws to remove excessive play.**

LEVEL THE TABLES. **Use a straightedge (commercial or shopmade) to check that tables are in the same plane. Metal shims can be inserted along the dovetailed ways to correct for tilt.**

SQUARE THE FENCE. **If the cast-iron fence is warped, attach a piece of plywood or MDF to the fence and shim it flat and square.**

SET THE KNIVES. **Knives must be parallel to the tables and set to the correct height. Shopmade magnetic holders can assist with the installation.**

Adjusting the Gibs

Over time the dovetailed ways may wear and cause one or both tables to go out of alignment. Tightening the gib screws removes slack and may correct the problem.

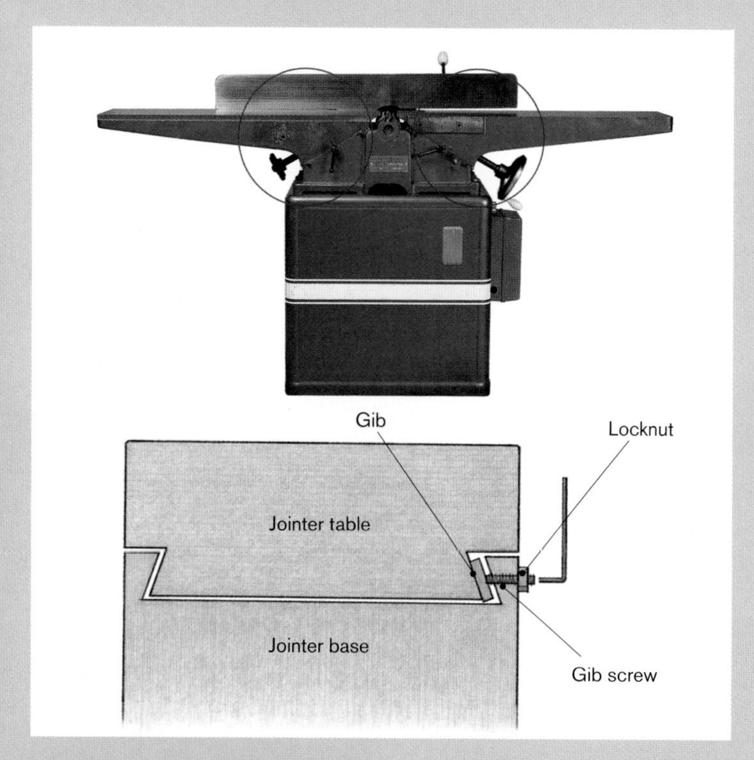

YOU NEED NOT DISASSEMBLE A JOINTER to do a tune-up. The narrow, flat bar is a gib, which takes up wear in the dovetailed ways of the infeed and outfeed tables.

THE OUTFEED AND INFEED TABLES have one gib each with two or three adjusting screws. Loosen the locknut and snug up the Allen-head screw to take up any slop. Tighten the screw nearest the cutterhead a tad more than the bottom one.

Unplug the tool before starting. It's also not a bad idea to tape the edges of the knives to protect both you and the knives. It's all too easy to brush a finger or tool across their exposed edges.

Remove Excess Play from the Tables

Each table has an adjustable gib to take up play as the dovetailed ways wear (see the drawing at left). Loose gibs can cause the tables to be out of line with one another. The gibs bear firmly against the dovetailed ways but must slide smoothly. When new, gibs are coated with grease. Over time the grease wears off. A regular shot of penetrating lubricant will keep things moving smoothly.

Each gib has a pair of gib screws that can be adjusted to take up play as the dovetailed ways wear. The screw nearest the cutterhead has to resist the lifting force caused by the weight of the table's overhang, and it should be adjusted tighter than the lower screw. The third screw on many machines has a handle that serves as a locking mechanism.

Start by backing off the locking lever and the locknuts on the gib screws. Then tighten all screws equally until the table is just locked in place, then back off each of the screws about a quarter-turn. At this point the tables should move with little resistance. Now slowly turn the gib screw nearest the cutterhead while moving the table up and down using the adjusting knob or lever. When the screw is properly adjusted, moving the table should require only moderate effort. Once this adjustment feels right, hold the screw against turning and tighten the locknut. Check and readjust, if needed.

Repeat the procedure with the lower gib screw, but apply slightly less pressure. If your machine has a center screw with a locknut, adjust it last and with only light pressure. Getting the gibs adjusted just right is a matter of both technique and feel, much like tuning a musical instrument. If you're lucky, the tables will now be aligned in a flat plane within 0.005 in. or less. Check them using the test bar or a long straightedge. If you have an older jointer, chances are that more will need to be done.

Tables Can Be Shimmed Level

Begin by removing the fence. Place a short straightedge across the cutterhead gap and lift both tables until they clear the knives and are in the same plane. Lock them in place. Lay a long straightedge or test bar across both tables. Use feeler gauges to measure any gaps (see the photo at left on p. 55).

On an older machine it's a good bet that the tables are sagging. To fix it, place thin metal shims along the dovetailed ways to shift one table into alignment with the other (see the right photos on p. 55). Flat shim stock may be purchased from machine-shop suppliers. Hobby shops also sell thin pieces of sheet brass and aluminum. Aluminum soda cans will also work; they are about 0.005 in. thick. Use a feeler gauge to measure how much the outfeed table is out at the far end. If you measure more than 0.005 in., the table should be shimmed. Anything less than that is probably not worth bothering with for the simple fact that you won't be able to find shim stock thin enough to make the fix.

Shimming is a trial-and-error process. As a rough guide, if your table is out by 0.006 in., start by cutting two pieces of 0.002-in.-thick shim stock that measures about 1 in. by 2 in. and apply a light coat of grease on them. To place the shims, back off the outfeed table's gib screws a turn, lift up on the low edge of the table, and slip the shims into place on the lower end of the dovetailed ways. Once the shims are in, adjust the gib screws again. Then realign both tables flat to one another and check for flatness using the long test bar or straightedge. The process may have to be repeated a few times.

Jointer tables may be tilted the other way and be dished. Follow the same procedure, but place shims at the upper ends of the dovetailed ways on the outfeed table. If you notice that the table is twisted, add thicker shims on the low side. Some small jointers may have a fixed outfeed table, in which case you have no choice but to shim the infeed table. Because the infeed table is adjusted frequently, shims may shift position or tear.

The Fix for a Crooked Fence

A small crown or dip over the length of the fence is tolerable as long as the fence remains vertical to the tables. A twist or wind, however, will give you fits, because it will cause stock to rotate as it passes by. To correct the problem, drill holes in the soft, cast-iron fence and attach a piece of cabinet-grade plywood or medium-density fiberboard (MDF) and shim it flat. Once the fence is mounted back on the jointer, square it up and take a test pass with a board that has a flat face. Check the resulting edge with a square, being sure to place the square against the board surface that ran along the fence. Adjust the fence stop as needed to get a square edge on the board.

Sometimes You have to Replace a Jointer's Knives

Nobody seems to enjoy replacing jointer knives. That's because it's difficult to keep the knives in alignment when tightening the bolts that are threaded into the lock bars (also called gibs). Patience is required, no doubt about it. Magnetic knife holders, either commercially bought or shopmade, can help.

First Find Top Dead Center Before replacing the knives, top dead center (TDC) of the cutterhead must be located (see photos 1-4 on p. 56). TDC is a point directly above the centerline of the cutterhead. When a knife's edge is at TDC, it is at the high point of its arc, the ideal spot to align it level with the outfeed table.

Sources

Esta
ESTA-USA, Inc.
P.O. Box 271
Barryville, NY 12719
800-557-8092
www.estausa.com

Remove One Knife at a Time Rotate the cutterhead until the edge of one knife is at TDC. Lock the cutterhead in place with a softwood wedge against the infeed table. Remove the knife and clean all of the parts, including the slot, of sawdust and pitch. Smooth the face of the locking bar and bolts with emery cloth or a stone to remove burrs, which may cause the knife to creep when tightened. It's important to remove and replace only one knife at a time to avoid distorting the cutterhead.

I do a lot of sharpening, but jointer knives take a lot of time and equipment to do well. I keep an extra set of knives on hand and send the old ones out to be resharpened after swapping them.

Adjust the Knives Cutterheads and knives come in various configurations (see the drawings on p. 56). Some cutterheads have springs beneath the knives. Better machines may have jackscrews in place of springs. Consider yourself lucky if you own a jointer with a cutterhead that accepts disposable knives, such as those made by Esta. With these, no depth adjustments need be made after the initial setup (see the top photo on p. 56).

Jackscrews allow the height of knives to be adjusted easily. Each knife rests on a pair of jackscrews that are set inside a hole in the cutterhead slot. Wedge the cutterhead at TDC, remove the first knife, install a fresh one, and snug up the bolts, leaving just enough slack for the knife to be moved without slop. Unless the owner's manual says differently, adjust the jackscrews until about 1/32 in. of the back of the knife (measured below the bevel) protrudes above the cutterhead slot.

Next, lay a short straightedge on the fence side of the outfeed table and extend it over the knife. Adjust the outfeed table until the knife just grazes the straightedge. Then place the straightedge along the rabbeting side of the outfeed table and adjust the other jackscrew, if necessary, to bring the knife into line. Rock the cutterhead back and forth; the knife should just kiss the straightedge. Tighten the bolts in progression to avoid warping the cutterhead, and check the setting again. Repeat for the other knives without changing the outfeed-table height.

On some machines the back edge of the cutterhead slot may be machined away so you cannot accurately measure from the back bevel of the knife to set the cutterhead height relative to the outfeed table. Instead, you have to set the outfeed table with a straightedge and feeler gauge 0.015 in. above the cutterhead (see the top left photo on p. 57). Make the measurement along the smooth surface of the cutterhead midway between slots.

On a cutterhead with springs instead of jackscrews, align a knife to TDC and wedge the cutterhead in place. Replace the dull knife, pressing it down against the springs. Snug up the bolts, leaving enough slack so that the knife may be moved but without slop. Place a pair of shopmade or commercial magnetic knife holders over the knife, which will lift it to the height of the outfeed table (see the bottom photo on p. 57). Lower the outfeed table until the back edge of the bevel on the knife drops below the outside surface of the cutterhead, then raise the table until the bevel and about 1/32 in. of the back edge of the knife protrude above the cutterhead. (Be sure to check your owner's manual on this matter.) Tighten the bolts in progression. Repeat for the other knives.

Leveling the Tables with a Shopmade Test Bar

I had hoped that a builder's level would be adequate for tuning up a jointer's tables, but I found it unfit for the task. Machinists use precision straightedges that are meant for just such applications, but at $200 for a 4-ft. version, woodworkers would have a hard time justifying such a purchase.

In search of a shopmade solution, I adapted a machinist's technique for creating precision squares. Technically, the resulting tool isn't a straightedge, because only the three slightly proud screws along one edge are in line. It is more properly called a test bar.

You'll need three bars of the same length and spacing of screws. The screws are adjusted by laying pairs of the bars flat with the screw heads touching. With each pairing, the height of the center screws is adjusted until all three sets of screws touch without either a gap or rocking. This process is repeated several times with different pairings of the bars until all three mate in any combination. When this is achieved, the laws of geometry dictate that the screw heads on each bar lie in a perfectly straight line.

Making a Test Bar

1. Rip three pieces of ¾-in. MDF, each about 5 in. wide and as long as your jointer.

2. Slope the ends of one board (A) to reduce its weight; it will become the test bar.

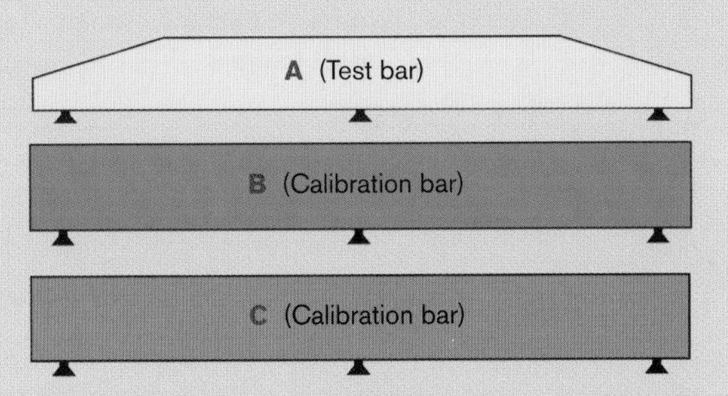

3. Next, predrill the edges of each board and attach three fine-thread, 1¼-in. drywall screws. Place two screws at the far end and one near the center of each board. File the head of each board. File the head of each screw to remove any burrs. Adjust them all so that ¼ in. of screw is exposed.

Adjusting a Test Bar

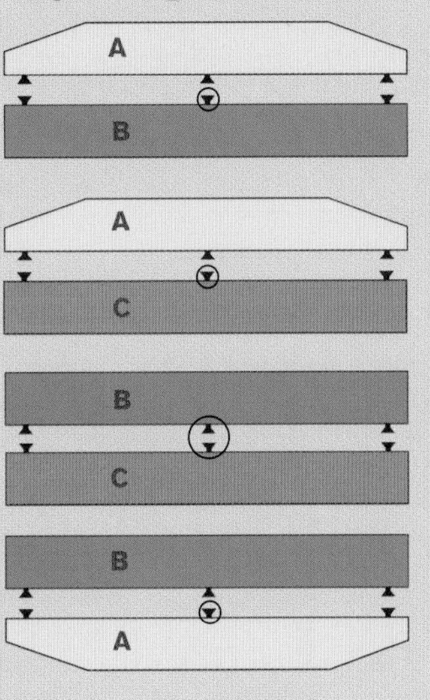

1. Align board A against board B. Adjust only the center screw on board B until all six screws touch.

2. Place board C against board A. Adjust only the center screw of board C until all six screws touch.

3. Place board B against board C. Adjust both center screws an equal amount (in or out) until all six screws touch.

4. Again place board B against board A, but this time adjust only the center screw of board A. Repeat steps 2 to 4 until no more adjustments are needed.

Making a Test Bar

USE THE TEST BAR AND FEELER GAUGES to check the tables for flatness. Tables may sag over time. New machines, however, may be out of adjustment, too.

IF THE OUTFEED TABLE SAGS, insert shims on each side of the lower section of the dovetailed ways. If the tables are dished (low in the center), shim the dovetailed ways near the cutterhead.

Setting the Knives

TYPES OF CUTTERHEADS

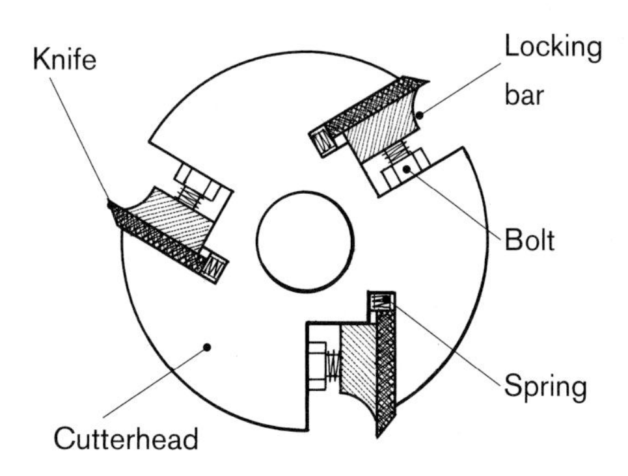

Some cutterheads use springs to adjust knives up or down. A clean shop floor will help ensure that you can track down an AWOL spring that will inevitably roll off the table.

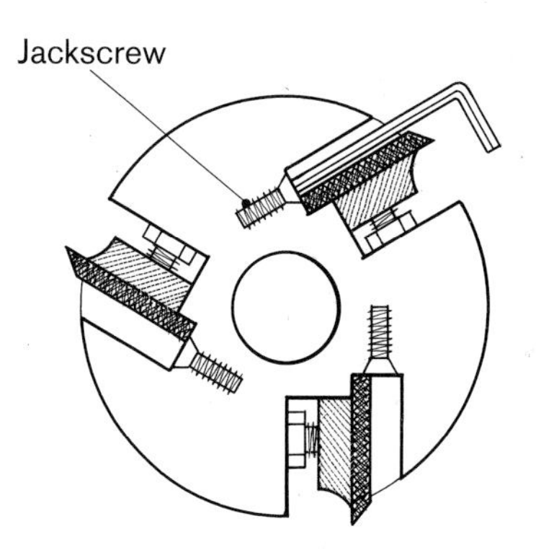

Some cutterheads use jackscrews to adjust knives up or down. The jackscrews fit into holes bored into the cutterhead slot.

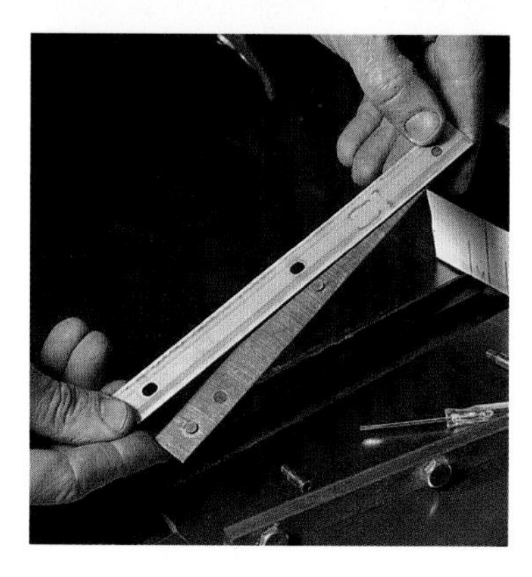

REPLACEABLE, DOUBLE-SIDED KNIVES MAKE LIFE EASY. Aftermarket kits such as this one made by Esta are available to fit most jointers.

Removing the Knives

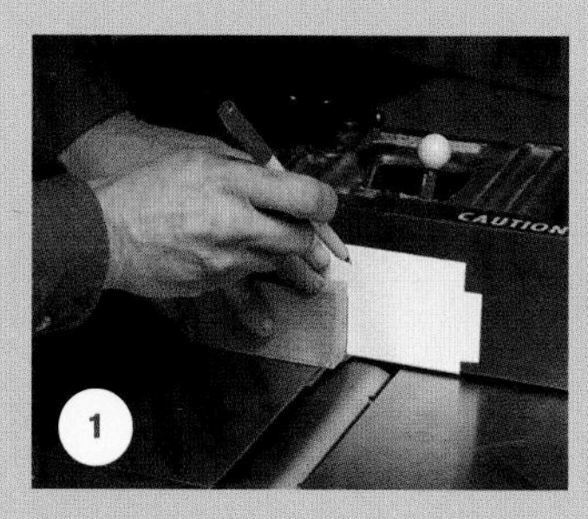

STEP 1

First find top dead center of the cutterhead: Slide a flat block of wood with a projecting screw head until the screw butts up against the cutterhead.

STEP 2

Mark this spot on the fence. Then do the same on the side.

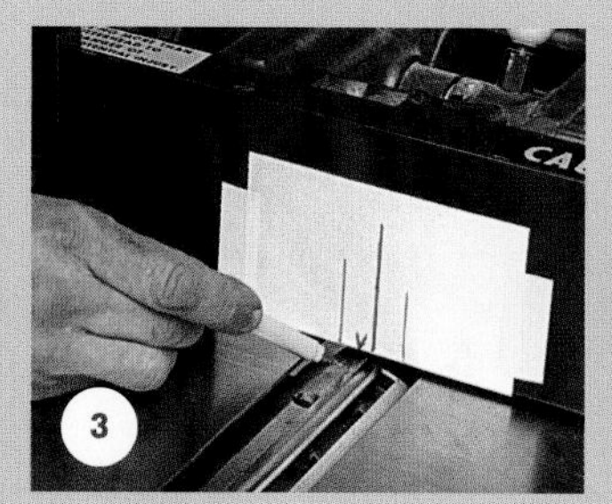

STEP 3

Using a ruler, split the distance between the marks to locate top dead center. Place another mark to indicate where the cutter-head slot lines up.

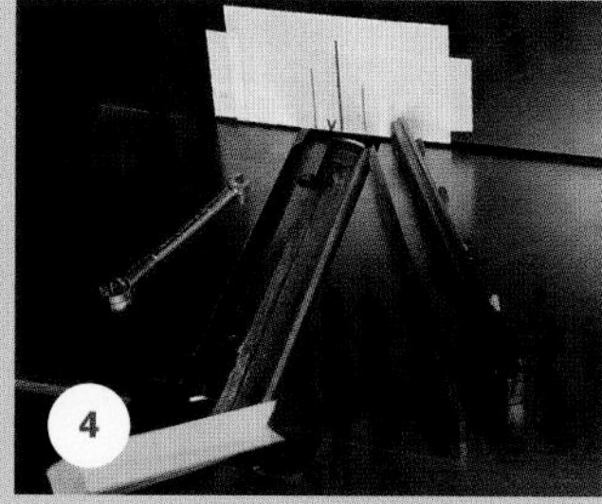

STEP 4

Align a knife to top dead center and wedge the cutterhead in place. Loosen the locking bar bolts and replace one knife at a time.

Make a Test Cut Before powering up the machine, double-check all of the bolts and screws that were adjusted. Then be sure to remove all tools from the machine. Spin the cutterhead one more time by hand to make sure that it moves freely and that no stray tools or parts have fallen into the slots.

Set the machine for a shallow pass and joint the edge of a reasonably straight board. If the end of the cut is deeper (snipe), raise the outfeed table slightly. If

Replacing the Knives

ON MANY JOINTERS, the outfeed table should be about 0.015 in. above the cutterhead.

ON OTHERS (SEE YOUR MANUAL), set the outfeed table so that it is level with a knife (at top dead center) when about 1⁄32 in. of metal behind the bevel is exposed.

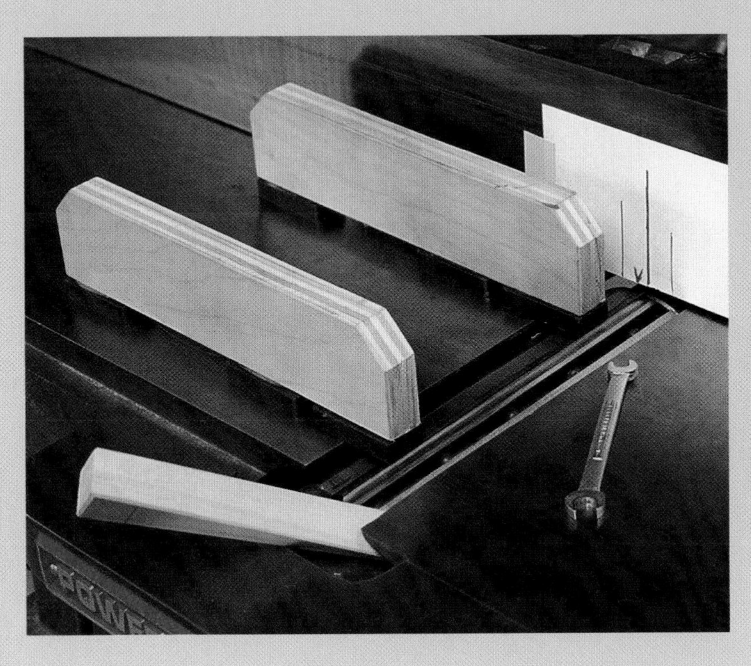

BAR MAGNETS (available from hardware or electronic stores) glued to a block of wood (do this on a level surface) with silicone adhesive make a decent knife holder/straightedge. Tighten the bolts on the locking bar so that each knife (at top dead center) is level with the outfeed table.

there's a hump of excess material left behind, lower the outfeed table. A lot of chatter marks mean one knife is slightly higher than the rest. Recheck the height of the knives. Aside from occasional knife replacement, the full tune-up may not have to be repeated for years and years to come.

JOHN WHITE is a woodworker and machinist who maintains the *Fine Woodworking* workshop.

Getting Peak Planer Performance

BY ROBERT M. VAUGHAN

When it comes to dimensioning stock, a thickness planer is indispensable. That is, unless the knives are dull or the machine's adjustments are out of whack. Dull knives are noisy and strain the motor. Nicked knives produce a molded surface instead of a flat one. Planer misadjustments cause end snipe, tearout, chatter marks, and feed difficulties. Improper planing technique also leads to poor surfacing. Until you are sure that your machine is adjusted properly, it's hard to tell whether your planing problems originate with the tool or with the user.

Fortunately, you don't have to be an experienced machinery mechanic to install knives or troubleshoot your planer. With a little patience and the right tools, you can diagnose and tune up your own machine (see the photo at left). To get predictable results, you'll need two gauging devices, which will let you observe measurements that you may otherwise gain only by trial and error and by feel. First use a gauge that rests on the cutterhead to set the knives. Then use a gauge that rests on the planer bed to measure the relationships between the cutterhead and the machine's other critical parts. These two gauging instruments, which have been used in the woodworking industry for at least 75 years, are simply dial indicators mounted in customized bases.

For up to several hundred dollars, you can buy gauges from various machinery manufacturers or aftermarket sources. But if you need to save your pennies for another

SIMPLE TOOLS AND INEXPENSIVE GAUGES can improve planing if you are patient and careful. Here, the author is installing knives in a 12-in. Parks planer using a dial indicator mounted in a base made to fit the cutterhead. Other tools for the job include a bed-resting gauge (by his knee), a mallet and block for tapping, and a prying tool (in hand). Note that the planer's plug is disconnected.

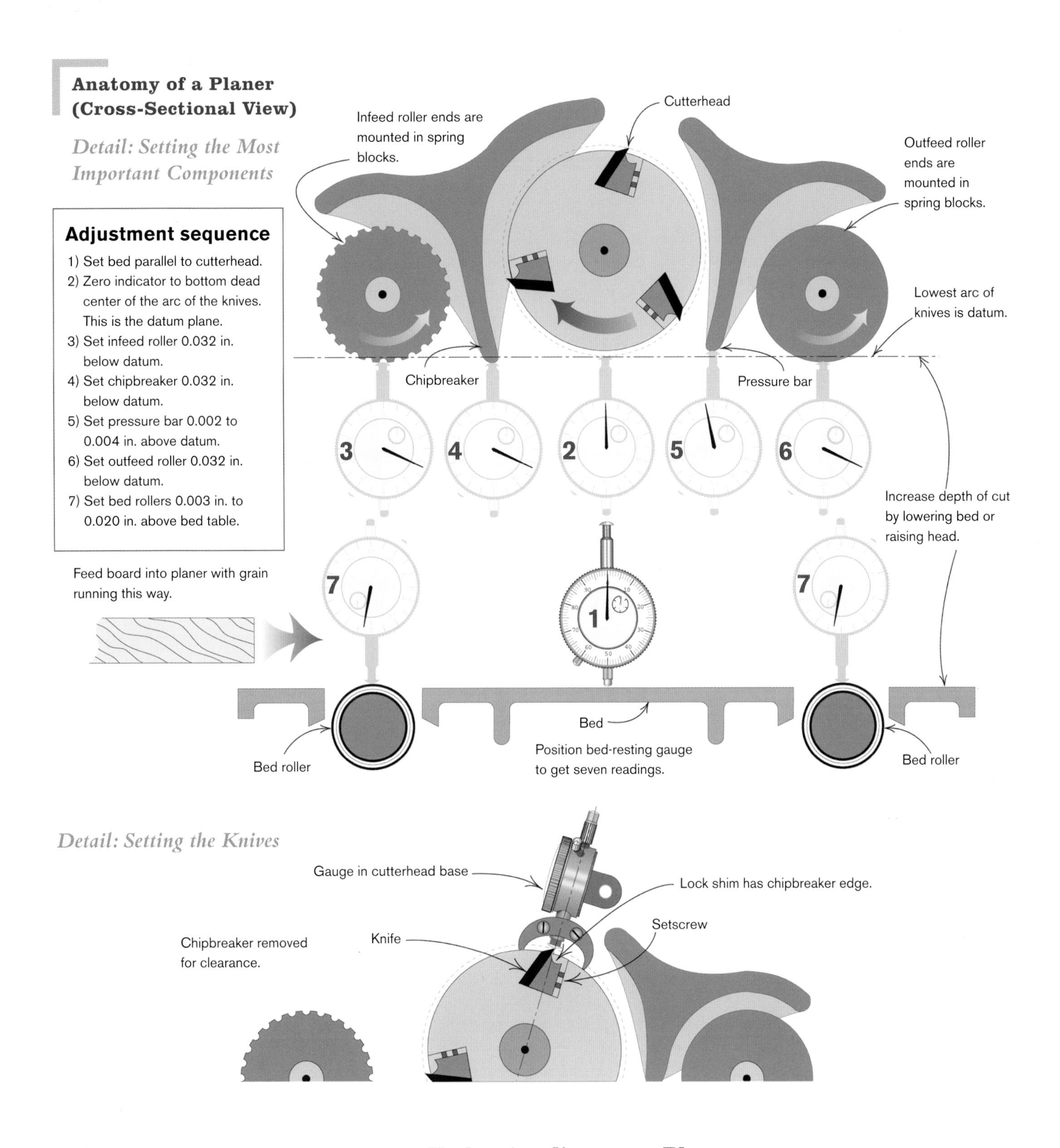

tool purchase, I'll show you how to make your own gauges using wood (or plastic, aluminum, or steel), a few nuts and bolts, and an ordinary dial indicator of the proper size with about a ¼-in. plunger range (see the sidebar on p. 61).

Understanding your Planer

The Parks planer shown in the photo on the facing page, though it is no longer made, contains all the common features found on a thickness planer. Your model may not contain all the components I'll address here. Even so, you should still be able to adapt the same principles to make adjustments to your own machine.

As a board passes through a planer, it is influenced by the relative positions of seven different components: the knives in the cutterhead (above the stock), the bed and bed rollers (below the stock), the infeed roller and chipbreaker (above the stock on the infeed side), and the outfeed roller and pressure bar (above the stock on the outfeed side). The drawing on p. 59 shows the relationships of these parts and the initial adjustment settings. Later, if I need to, I'll tweak with the adjustments to fine-tune the planer's cut.

To understand where each of the planer's seven components plays its role, it's helpful to follow a board as it's being planed. First the wood is placed on the planer bed and fed by hand between the infeed roller and the front bed roller. The powered infeed roller grabs the wood and drives it beneath the floating chipbreaker and under the rotating cutterhead. Next the board passes under the pressure bar and out between the powered outfeed roller and the back bed roller as it exits the machine. Having any of these components out of whack will cause problems, so checking each is essential. Start by setting the knives in the cutterhead. But before you do anything, prepare the machine, and get the tools you'll need.

Preparation

First, unplug the machine. You'll also want to disconnect the dust boot to gain better access. Then remove the guard for the pulley, so you can advance the cutterhead. Besides the dial-indicator gauges, you will need a few other tools: an ice pick (or other device to pry up the knives), a wooden block and a mallet to tap the knives down,

Assembling the Gauges

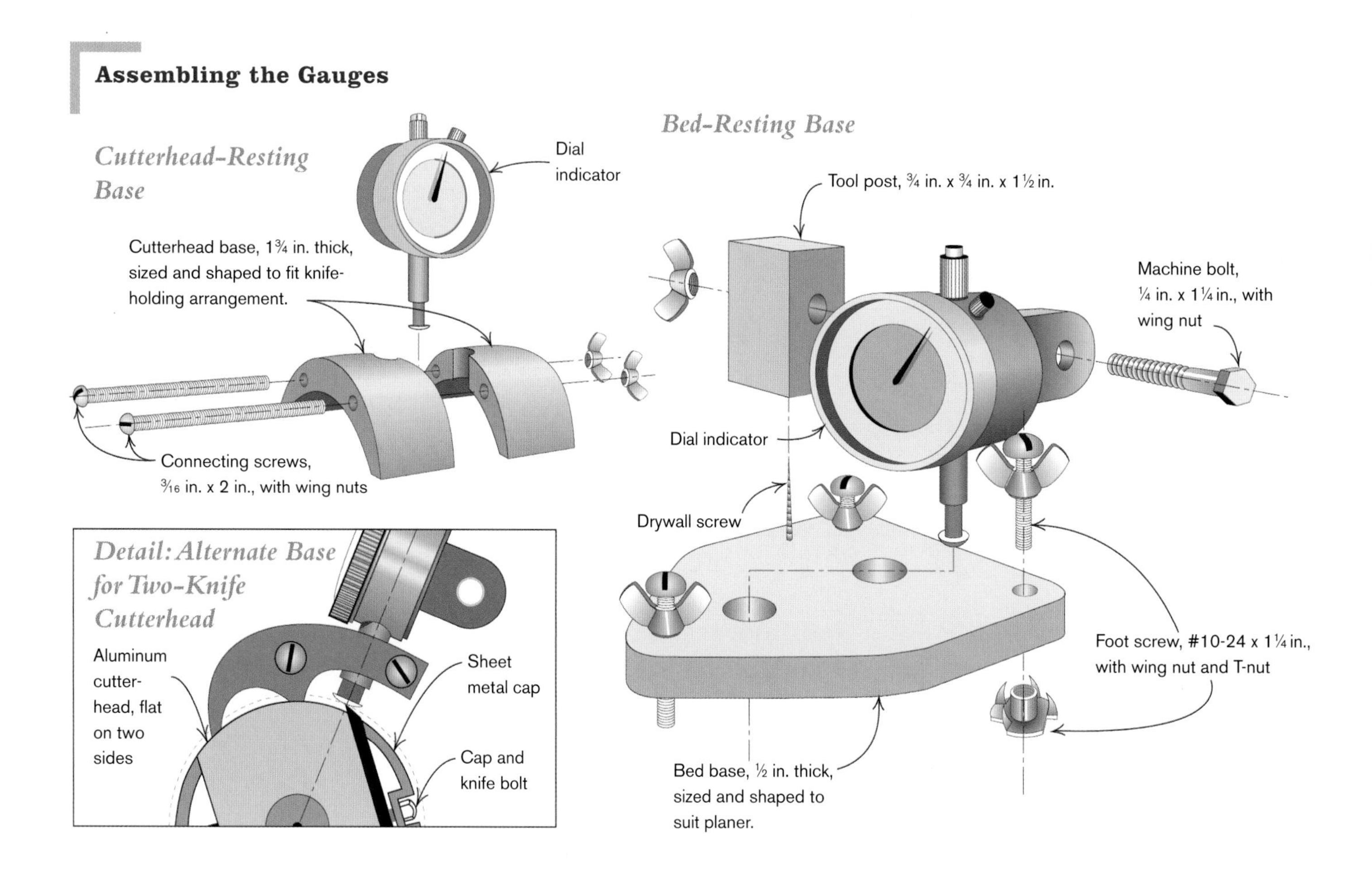

and Allen wrenches to tighten the lock shim and to turn the jackscrews (if your machine has them). Study your owner's manual so that you will know how to adjust the components on your particular machine and gather the required wrenches. Some metal shims may be handy for fine-tuning adjustments. Depending on what you find once you get into the job, you may also need a file, some emery cloth, and solvent and lubricant. And make sure you're comfortably seated.

Setting the Knives

When setting the knives parallel to the cutterhead, remove and reset one knife at a time to avoid distorting the head. This requires a spare sharpened set of knives. I always have my knives sharpened at a professional sharpening shop. If knives are being installed on an empty cutterhead, then lightly install all the knives, and go from knife to knife, gradually increasing pressure. For maximum support and safety, the knife should be as far down in the slot as practical. There may be differences between cutterheads, too. Some have jackscrews, or there may be two knives in the cutterhead instead of three (see the drawing detail on p. 59).

To check if your old planer knife is a safe size after resharpening, remove the lock shim (also called a lock bar or gib) from the cutterhead, and lay the knife about where it should be. If you see any light through the setscrew holes, reject the knife; it is too narrow and could be thrown from the cutterhead. Don't exert a lot of force on the setscrews, or you'll distort the cutterhead and the screw threads. Apply equal torque on the screws to get uniform pressures and deflections. I get enough leverage from the 6-in.-long leg of my Allen wrench. It's a good idea to lay a rag over the exposed blade to protect your hands in case you lose your grip.

Shopmade Planer-Setting Gauges

My shopmade gauges were adapted from the heavy steel gauges I service planers with in the field. For occasional use, the shopmade gauges give equally precise readings. I devised the gauges so one dial indicator can be interchanged from one base to the other. Because planer dimensions vary, the bases' measurements will also vary. To size them, first get the right dial indicator.

Selecting a dial indicator: A dial indicator is excellent for showing crucial relationships of machine components. One of the inexpensive imported units goes for about $25 (Enco Manufacturing, 5100 W. Bloomingdale Ave., Chicago, IL 60639; 800-873-3626). After you thoroughly study the parts of your planer and all its adjustment limits, sketch a full-scale cross section of these (see the top drawing detail on p. 59). This will help you choose a dial size and also show you how the bases need to be shaped. Select an indicator that will fit easily and can be read clearly in your planer. I use a 1¾-in.-dia. dial with a ⅜-in.-dia. convex replacement tip, like Starrett's or Mitutoyo's hardened, chrome-plated type. Convex tips provide better contact over a knife.

Making the bases: To make the cutterhead-resting base, first make a full-size sketch of your cutterhead (see the bottom drawing detail on p. 59). Extend a line from the center of the cylinder out over the tip of a knife. Position your indicator over the line with the plunger pointing at the center of the cutterhead. Next draw a base profile with two feet resting on the cutterhead. For two-knife cutterheads, try making an indicator base that has both feet on one side (see the drawing detail on p. 60). Mark where the plunger stem passes through the base. Then transfer your base profile to a block of wood, and drill and cut to size. Using the hardware shown, assemble the gauge.

For the bed-resting base, make a crow's-foot (tripod) arrangement. The position of the screw feet should be such that the feet won't drop down in the bed-roller slots. Use the planer sketch to locate the indicator tool post. I devised mine so that I can swap the indicator from the front to the back of the post. Finally, round and polish the bottoms of the base's screw feet.

–R.V.

To use the cutterhead gauge, I lightly tighten a knife close to its proper height. This varies from machine to machine, so you should check your owner's manual for the recommended height. I then clamp the dial indicator's ⅜-in.-dia. shaft in the

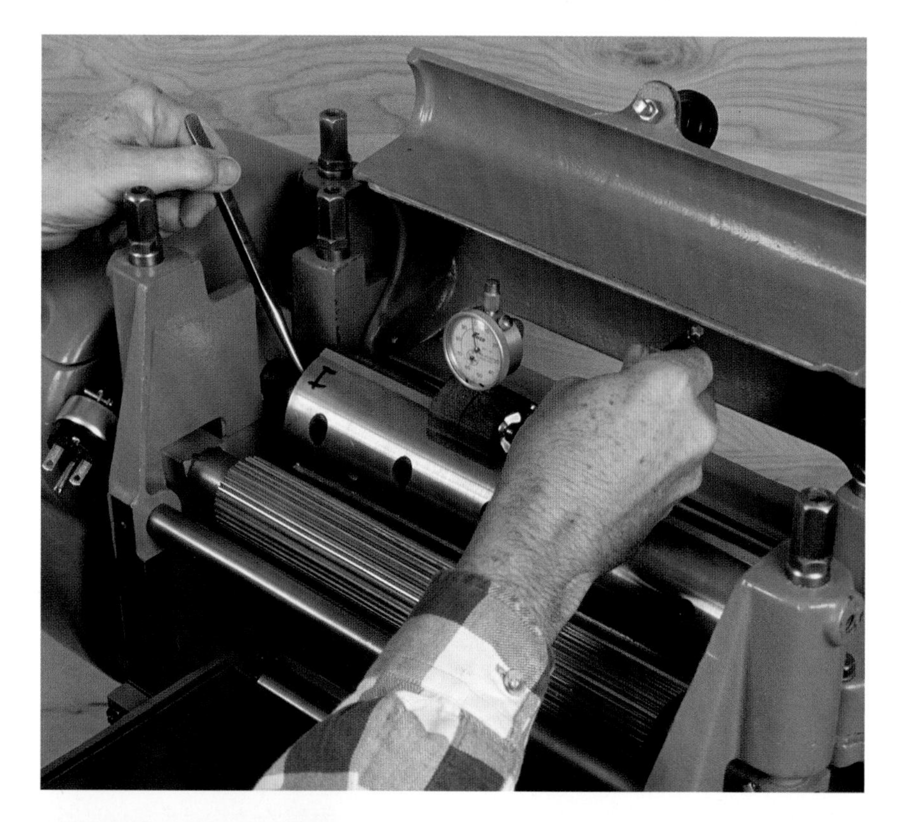

GAUGE HELPS TO ANTICIPATE KNIFE SHIFT. **Using the planer cutterhead as a reference, Vaughan reads the gauge over each setscrew to know whether to raise or lower the knives and to anticipate how much each of the knives will shift during tightening.**

A CUTTERHEAD GAUGE ENABLES **knives to be set consistently to within one or two thousandths of an inch. The wing nuts on the base allow plunger height adjustment.**

wooden base so that when the base is rocked on the cutterhead, the dial will move only about 0.015 to 0.020 in. At this point, I turn and lock the moveable dial face, so the indicator's hand points to zero when the plunger tip is moved over the tip of the knife (see the photo at left). Now the indicator will register the height of the knife edge relative to the cutterhead.

Lightly tighten the setscrews on the outer ends. I usually snug the left side to exact position, go back to the right side and raise or lower that side of the knife to where it should be, and lock it in position. Then working from left to right, over each setscrew, I either raise or lower the knife until I can lock it at the proper height (see the photo above). Rocking the indicator's plunger over the knife edge shows me the maximum protrusion of the knife edge. Keep in mind that the wood of the gauge base is light and sensitive. Take a few minutes to get the correct feel of the gauge base contacting the round cutterhead. Repeat the sequence—one knife at a time—for the other knives in the cutterhead.

Setting the Machine

Once the knives are set, install the indicator in the other base with the plunger tip up. Drop the planer's bed until the bed-resting gauge can be easily placed directly beneath the cutterhead. Crank up the bed until the plunger tip just touches the bottom of the cutterhead. Be sure the cutterhead has been rotated so that the knives are out of the way. Then place the gauge at one end of the cutterhead, and rock the cutterhead as you zero the dial at bottom dead center. Zero the other end of the cutterhead as well. Brush the plunger under the center of the cutterhead. If there is a sizable difference (more than 0.015 in.) between the middle reading and the ones taken from the outsides, then the bed has been worn too much and needs to be remachined.

The Bed The planer bed and cutterhead should be parallel. How to make them parallel varies from machine to machine. Some machines require the table be adjusted and others require the head position be adjusted. For those machines that have no adjustment, the only option is to set the knives in the cutterhead, so they will be parallel to the bed instead of the cutterhead.

Before working with the dial indicator, make sure that the bed has no slop in it as it moves up and down. Most machines have wear shims that can be adjusted. A sloppily fitting head or bed will give poor surface results, such as snipe and washboard.

Defining the Cutting Arc Using the cutterhead gauge again, double-check (over each setscrew) the positions of the knives in the cutterhead. Final setscrew tightening often causes the knife to squirm up a hair.

Then position the indicator back in the bed-resting base so that the plunger is at bottom dead center of the cutterhead. Lower the table without disturbing the position of the gauge base. Rotate the cutterhead by hand until one of the knives is at bottom dead center. Carefully raise the table until the plunger tip just touches the knife. Reach in and steady the position of the base while raising the bed just enough to make the knife move the plunger about 0.015 in. Zero the dial when the knife rotates through bottom dead center of its arc (see the photo on p. 64). This defines on your gauge the lowest point of the cutting circle. This will be your datum. It is this plane that defines the position of the planer's upper internal components. Neither the bed nor the cutterhead positions should be disturbed while making the rest of the upper adjustments on the planer.

The Infeed Roller Straddle the bed roller slot with the feet of the gauge base, and move the indicator in and out under the infeed roller. The position for serrated steel infeed rollers should be about 0.030 in. to 0.035 in. below the cutting arc for most machines. Rubber rollers will be slightly lower. For sectional infeed rollers or chipbreakers, you'll have to average the measurements. Consult your manufacturer's literature to get an exact figure of the correct position in relation to the cutting arc.

When adjusting the infeed roller to the correct position, the face of the indicator may not be in the most convenient spot for viewing. If this is the case, cut a triangular block of wood about 2 in. high, and fasten a mirror to it with double-faced tape to view the results when standing above the planer. This mirror can be used for the other internal components as well.

The Chipbreaker Like the leading edge of a handplane's cap iron, the chipbreaker in a planer prevents long tearouts from occurring. The chipbreaker is often, but not always, set to the same distance below the cutting arc as the infeed roller. Proper alignment keeps long strips of wood from lifting as the top of the board is being cut by the knives. Set the chipbreaker to manufacturer's specifications using the gauge in the same way it was used to set the infeed roller. Some machines have antikickback fingers or pawls just ahead of the chipbreaker.

The Pressure Bar The pressure bar is located behind the cutterhead and keeps the newly cut surface from bouncing up into the cutterhead as the stock enters and exits the planer's feeding system. During the cut, it performs a hold-down function when feeding warped stock. If it is set too high, the wood will flutter and a washboard texture will result. And it's likely that end snipe (a slightly thinner section) will occur. If it is set too low, feeding will be impeded. A majority of surfacing problems can be traced to this component, so its position is critical. I normally set a pressure bar about 0.002 in. to 0.004 in. above the cutting arc for surfacing face-jointed lumber. For surfacing lumber that is rough on two sides, a slightly higher setting usually works well.

The Outfeed Roller The outfeed roller is usually smooth or rubber-coated, so it won't mark the planed surface. Set the outfeed roller exactly like the infeed roller. It should also be set to the same distance below the cutting circle, unless the manufacturer's instructions state otherwise.

The Bed Rollers The bed rollers reduce friction as stock is being fed, and they prevent premature wearing of the bed tables. So, it's important that the rollers turn easily and are aligned precisely. Bed rollers are located in slots in the bed directly below the two feed rollers. The dial indicator will have to be reinstalled in the base with the plunger down to check the position of the bed rollers (see the photo above). Adjust the feet so the plunger moves up only about

THE PLANER BED GAUGE has to work in different positions. The base's screw feet are located so the gauge can straddle the bed slots. To check the feed rollers, orient the dial indicator so it can take overhead readings. To check the bed rollers (shown here), flip the indicator on the tool post, sticking the plunger down through the hole in the base.

0.015 in. when the base sits on the bed. The weight of a wooden base is often not enough to overcome the opposing spring pressure of the indicator's plunger, so hold the base down for accurate readings.

The rougher the lumber, the higher the bed rollers should be set to reduce friction. However, if they're set too high, the workpiece may vibrate, producing a rippled surface. Conventional practice is to set the rollers 0.002 in. above the bed when dressing faced lumber and about 0.020 in. when dressing lumber that's rough on both sides. On this machine, I set the rollers to 0.002 in. and then insert 0.020-in. shims on those rare occasions when I'm dressing lumber that's rough on two sides (see the top photo on the facing page). Machines with no bed rollers don't usually have performance problems related to the lack of bed rollers. But the beds don't stay flat nearly as long either, and the motors work a bit harder.

THE BED GAUGE CHECKS **infeed and outfeed components. It also shows if bed adjustments are needed. By turning the pulley, the author rocks the cutterhead to be sure that he is reading bottom dead center of the arc of the knives.**

Helpful Hints to Better Surfacing

The dimensions I have shown are those I use for a starting point when adjusting planers and are far from being written in stone. Other factors such as component wear, wood dryness, wood straightness, and operator preferences can easily dictate that things be adjusted differently.

Adjustment Problems Adjustment screws on planers usually are held in place with locknuts. When the correct settings are reached by turning the adjustment screw, those settings usually alter when the locknut is tightened. It's always a good idea to watch the indicator's hand when the locknut is tightened, so the setscrew can be turned to compensate for the difference.

Spring Pressure Downward spring pressure can sometimes have an effect on planer performance. A heavy spring can emboss infeed roller prints on softwood when making that light final pass. Light pressure can cause roller skidding when rough or warped lumber is dressed. How much is enough? Only the performance of your machine will tell you that.

Safety Because planers pull the wood away from you, loose clothing and jewelry can be a hazard. Noise is also a factor. When knives get dull, they loudly beat off the chips rather than cut them. So always wear ear plugs in addition to eye and breathing protection. Try to cut out defects such as knots beforehand, and never plane a board that's less than ¼ in. thick or shorter than the distance between the feed rollers.

Any cutterhead that is moderately exposed on the outfeed side should have a shroud over it to prevent easy access to the spinning knives. Drive belts and gears should also be covered, so you don't come in contact with such moving parts. Last, never look into the machine (infeed or outfeed end) when it's running.

TEMPORARY BED-ROLLER SHIMS **make heavy milling easier. When Vaughan wants to do heavy planing, he elevates the bed rollers with temporary shims. This is easier than having to adjust each end of both rollers individually. The shims, tethered on a string for convenience, are removed when it's time to do finer surfacing.**

Dust Collection Though this machine was not shown with a dust collector, for best planing results, as well as for health concerns, you should have a dust- and chip-evacuation system. Chips can pile up and get pressed into the wood under the outfeed rollers and get dragged around by the knives. This makes for little dents on the wood that will eventually spring back as little bumps when the wood takes in more moisture. Ideally, your planer should produce long, clean shavings (see the photo below).

Planing for Success Planer-operator technique can have as much to do with poor surface quality as a poorly adjusted machine. For example, slower feed rates tend to produce smoother surfaces. And hardwoods generally should be fed slower than softwoods. Also, keep these guidelines in mind when you are planing: Not supporting long stock as it enters and exits the planer will almost always result in a snipe. Trying to surface warped stock will usually cause a washboard surface because the wood is not flat on the planer bed. Taking too heavy of a cut can cause tearout; feeding the wood against the grain will cause tearout; and dressing knotty or highly figured wood increases the risk of tearout. Not taking a light final pass to get to finished dimension can result in a rough surface. High moisture content in the lumber makes the fibers stringy and difficult to cut cleanly. The result is a fuzzy surface. It also will likely be a bear to feed properly. Finally, a planer smooths stock and makes the faces parallel. It will not straighten warped stock.

ROBERT M. VAUGHAN is a contributing editor to *Fine Woodworking* magazine. He repairs and restores woodworking machines in Roanoke, Virginia.

THE PROOF OF PROPER PLANING **is all in the shavings. When a planer is producing long, clean shavings and little dust, chances are good that the machine is well-tuned. The smooth, tearout-free knot on this planed board of poplar shows that it's well worth the fuss of using dial-indicator gauges to install knives and set planer components.**

The Jointer and Planer Are a Team

BY GARY ROGOWSKI

JOINTER

The jointer has two jobs: It mills a single face of a board flat and straight, and it can square one edge to that face.

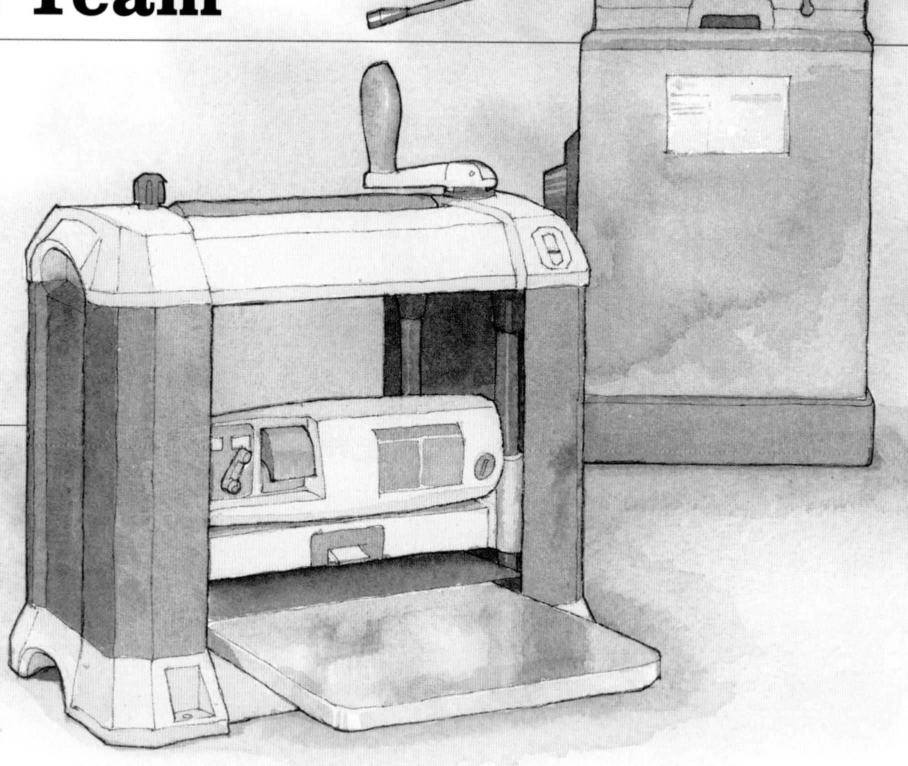

PLANER

The planer is better described as a stock thicknesser. Its job is to plane one face parallel to another.

My beginning students often ask me, "Which machine should I buy first, a planer or a jointer?" The answer is both. With a jointer alone, you can't get boards of consistent thickness. And with only a planer, you'll get consistent thickness, but your boards still can come out twisted or bowed.

Perhaps because of these machines' confusing names, many woodworkers don't grasp the separate functions they serve. The European names for these tools—planer (for jointer) and thicknesser (for planer)—are more accurate. The jointer planes a level surface, and the planer simply creates uniform thickness. Because of its American name, some woodworkers think the jointer is only for milling the edges of boards before glue-up.

Together, the two machines form the gateway to serious woodworking, allowing you to mill your own lumber to custom thicknesses instead of being stuck with the surfaced hardwoods available at the local home center. They also allow you to work with rough lumber, which is much less expensive than S2S (surfaced two sides) or S4S stock. Add a bandsaw or tablesaw, and you have the ability to dimension lumber to any width, thickness, and length.

The Jointer Comes First

This machine planes a flat face on a rough board, using the freshly planed section as a reference surface for the rest of the cut.

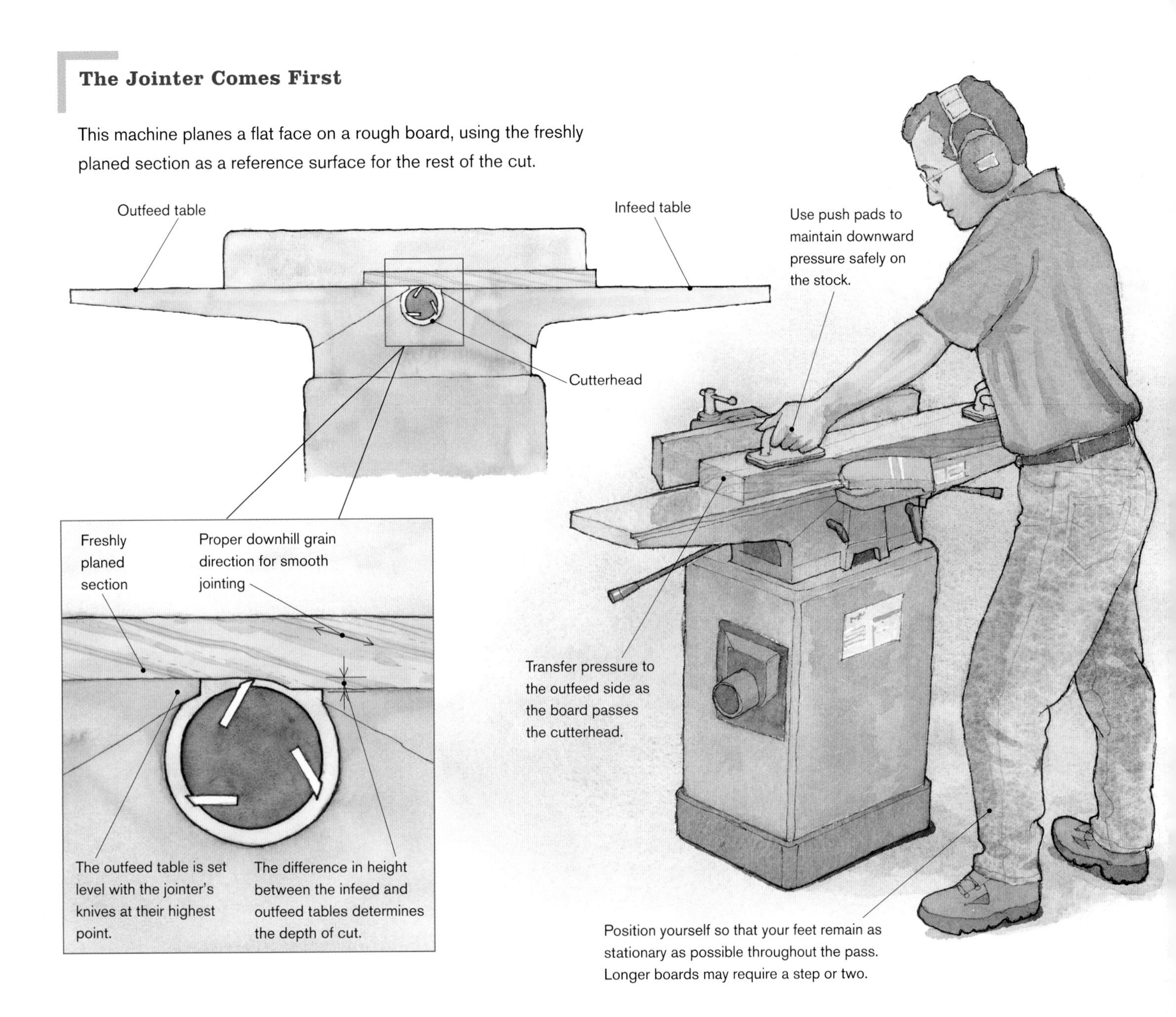

Which Face to Joint?

Chances are the lumber you are milling will not be flat. Orient the board so that the cupped or bowed side is down to prevent rocking during jointing.

Bowed board

Cupped board

With the bowed or cupped side down, the board rests steady on two points. The flats get wider with each pass until the surface is flat.

Rough-Cut Stock to Size Before Milling

If you need smaller pieces from a long, bowed board, cutting the board to rough length first will result in thicker stock. The same goes for width.

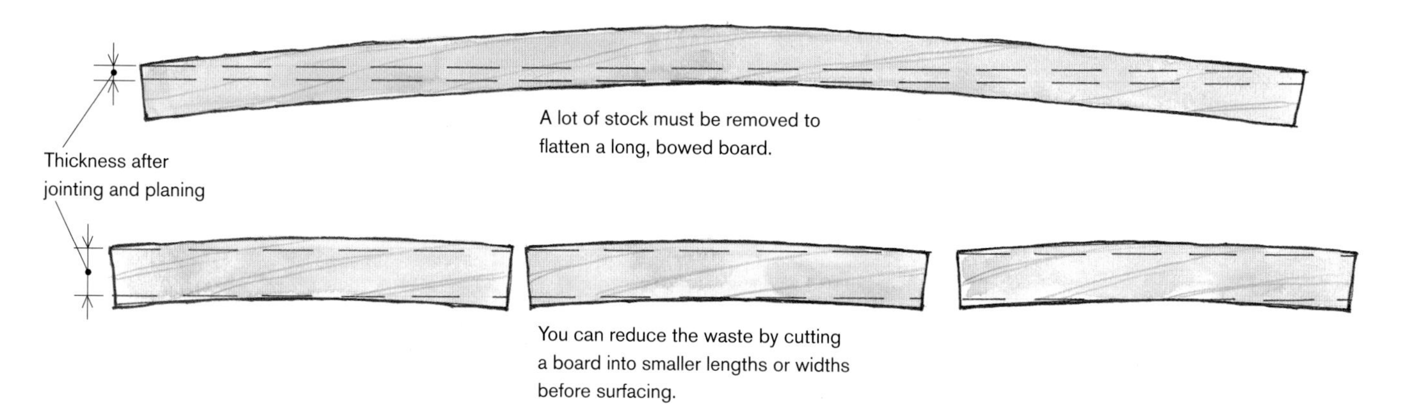

Thicknessing Starts on the Jointer

A jointer works like a handplane turned upside down, with its reference surfaces in line with its cutter knives. Use this tool for flattening one face of a board. If you flip over the board and joint the other side, there is no guarantee the faces will be parallel. On the jointer, each face is cut without referencing the other.

Start by Roughing Stock to Size Before jointing the first face, get your material roughed out to length and width. If a long or wide board is badly cupped or bowed, running it over a jointer until it's flat will waste a lot of wood. You also can rough out around board defects such as knots, sapwood, or checks. Use a chopsaw or handsaw to rough the stock to length, removing any checked or cracked areas on the ends. Next, rough your stock to width. This can be done in a variety of ways. If the board is badly crooked, you may need to snap a chalkline on it and bandsaw to the line. Otherwise, run one edge over the jointer or handplane the edge to level it out. Now you can rip the board to rough width.

I highly recommend a bandsaw for ripping rough lumber. It wastes less wood and is much safer because there is no danger of kickback.

Put the Cupped or Bowed Side Facedown It's highly unusual to find perfectly flat stock. That's because wood at a retail lumberyard gets uneven exposure to the air. Here's what to look for: cupping across the width, bowing along the length, and twist or wind in a board's thickness. First, check to see whether the board is cupped across its faces. Use a straightedge or check with your one good eye. It will be easier to run the cupped side down on the jointer table because the board will reference off its two outer edges and not rock. Take off small amounts of wood with each pass until you cut across the entire face and length of the board. Use push sticks or pads to hold the board firmly and safely on the jointer table. Mark the unjointed face with an X.

Twisted wood is deceiving. Use winding sticks to check your lumber or hold a board flat on the jointer table and see if it rocks when you push down on a corner. Mark the high corners of one face. On the jointer, start with all of your hand pressure on the leading high corner. As you continue the cut, transfer the pressure to the opposite high corner, trying to prevent rocking to one side or the other. Make multiple passes until the board is flat.

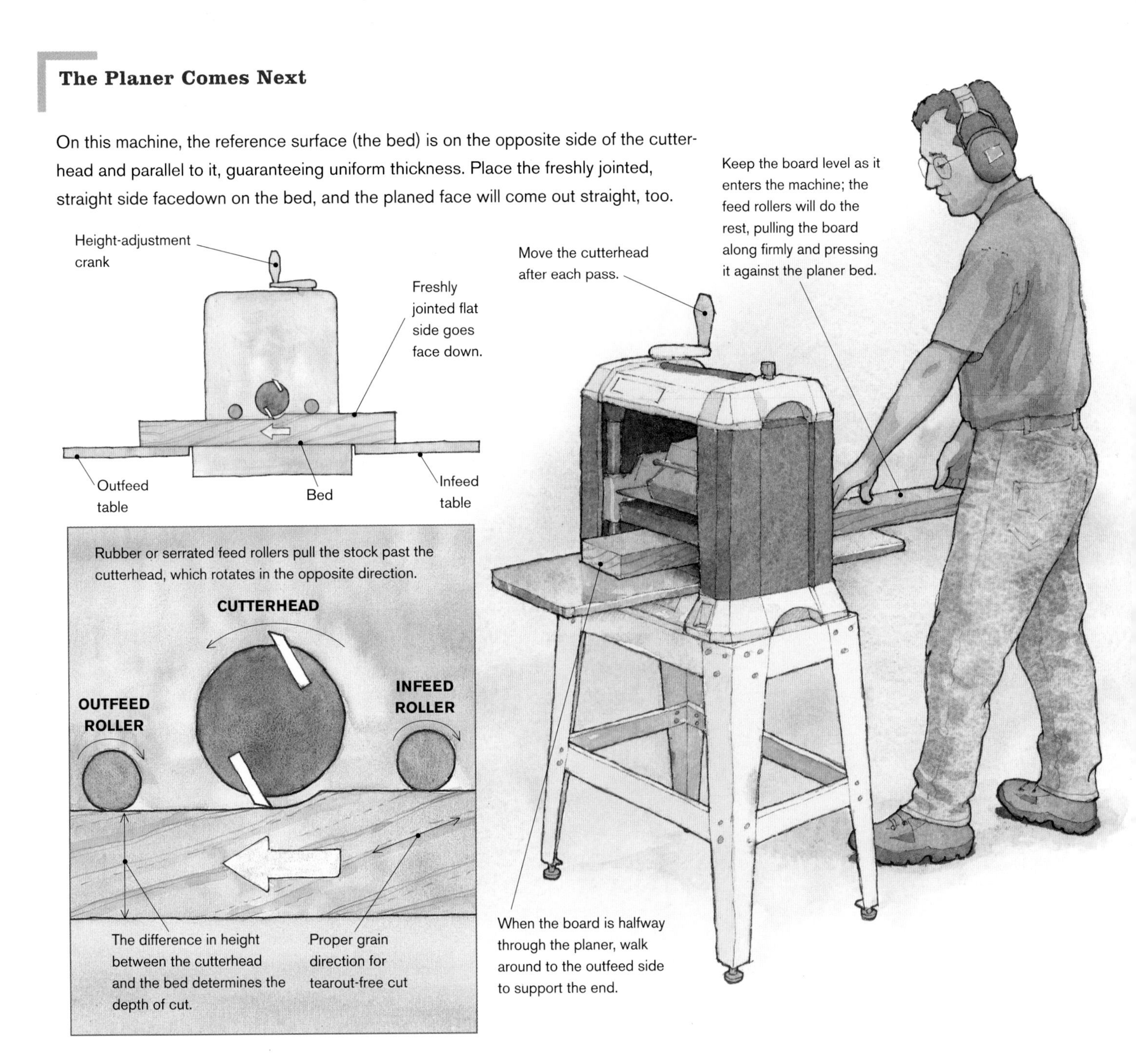

For any of these cuts, check the grain direction of the board before passing it over the jointer. And always keep your feed rate slow, use push pads for protection and to dampen vibration, and take shallow cuts.

The Planer Comes Next

The impatient woodworkers among you may think, let's skip all this bother on the jointer and go straight to the planer. Sorry, it won't work. The planer will take whatever bowed or twisted surface you give it and make a cut parallel to that face. The reference surface on a planer is the bed; the knives are above the stock. So if the board is bowed when it goes in, it will be bowed when it comes out. If it's cupped, the planer's feed rollers may flatten the board slightly, but when it comes out it will pop back to being cupped.

You must use the jointer first to flatten one face. Then run this straight, flat side face down in the planer to create a parallel, flat face on the other side of the board.

Arrange all of your boards for grain direction before starting the planer; remember, you're cutting on top of the board now. Make the first pass a light cut. If possible, feed the boards continuously one after the other, end to end, which eliminates the planer's tendency to snipe at the beginning and end of a board. Plane all of the boards down to thickness, leaving them a hair oversize to allow for removing the milling marks. These marks are not a decorative effect.

If you get tearout on a face no matter how you feed the board, dampen a rag and lightly wet down the surface of the wood before planing. This will help soften the fibers and tone down most of the tearout. Also, wax your planer tables.

Last, Mill the Stock to Width and Length

After your faces are flat and parallel, work on the edges. Check that your jointer fence is set square to the table just beyond the cutterhead on the outfeed table. This is where your hand pressure should concentrate once the cut is established. Check for crook along each board's edge, and run the crooked edge down to the jointer table. Mark the squared edge and face after cutting.

Rip the last edge to width on the tablesaw or bandsaw. If the cut is rough, you'll want to leave a little extra for one final pass on the jointer. Last, cut the ends to length. Crosscut one end square on all of your boards, using your crosscut sled or miter gauge on the tablesaw. Then clamp on a stop to index the remaining cuts.

GARY ROGOWSKI is a contributing editor to *Fine Woodworking*. He runs The Northwest Woodworking Studio, a school in Portland, Oregon, and is the author of *The Complete Illustrated Guide to Joinery* (The Taunton Press, 2001).

Why One Machine Is Not Enough

Jointers and planers are great at doing the jobs they were designed for, but you can get into trouble by asking them to do too much.

FLAT BUT NOT PARALLEL

If you use a jointer first to plane one side and then the other, you may end up with flat sides but an uneven thickness.

PARALLEL BUT NOT FLAT

If you use a planer to flatten the first face of a bowed or cupped board, it simply will follow the curve.

All About Bandsaw Blades

BY LONNIE BIRD

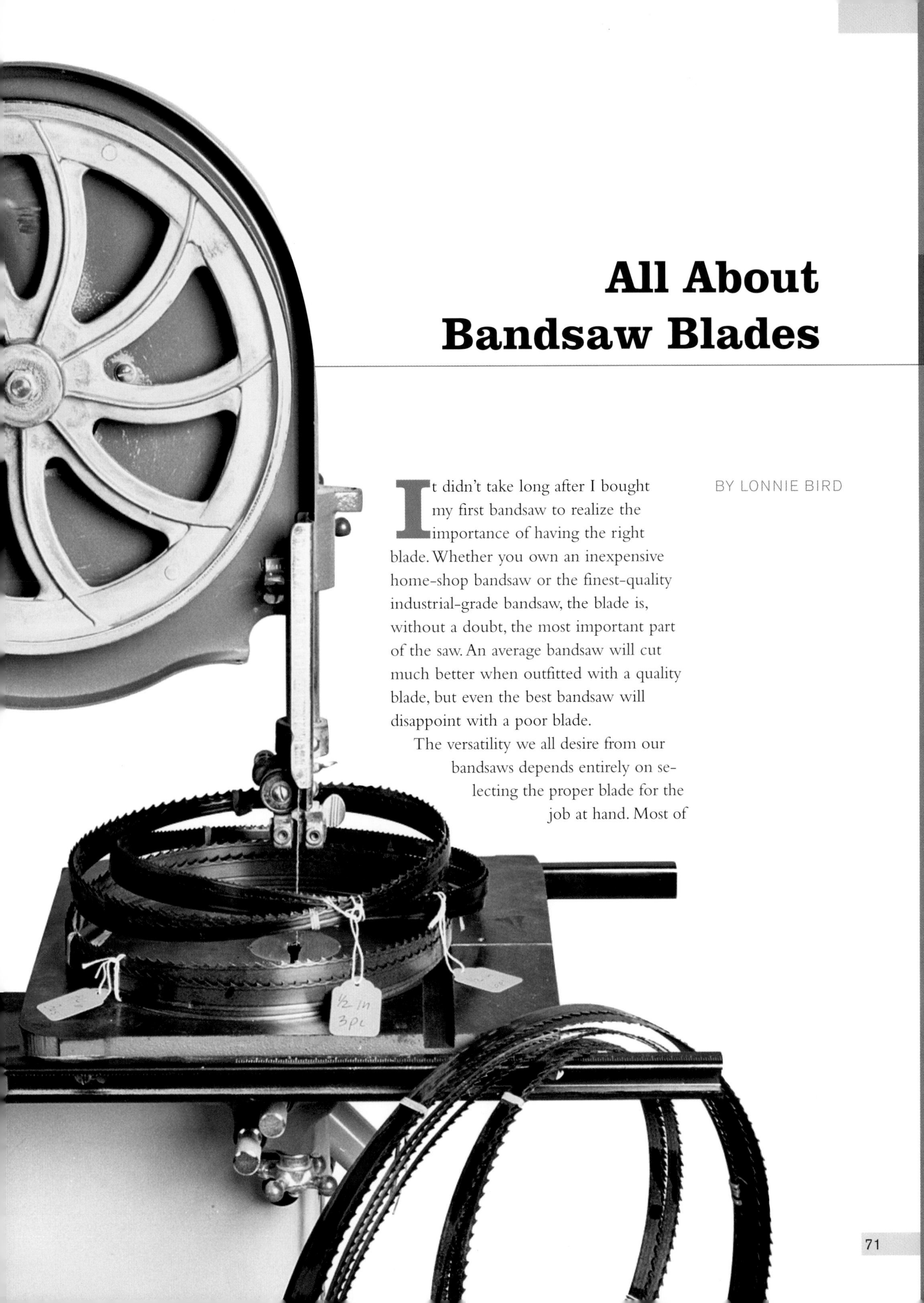

It didn't take long after I bought my first bandsaw to realize the importance of having the right blade. Whether you own an inexpensive home-shop bandsaw or the finest-quality industrial-grade bandsaw, the blade is, without a doubt, the most important part of the saw. An average bandsaw will cut much better when outfitted with a quality blade, but even the best bandsaw will disappoint with a poor blade.

The versatility we all desire from our bandsaws depends entirely on selecting the proper blade for the job at hand. Most of

Terms You Need to Know

BLADEBACK - The body of the blade not including the teeth. The bladeback must be both tough and pliable to withstand the continuous flexing as the blade runs around the wheels of the saw.

GULLET - The curved area at the base of the tooth that carries away the sawdust. The size and efficiency of the gullets decrease as the pitch is increased.

PITCH - The number of teeth per inch (tpi) as measured from the tips of the teeth. The pitch determines the feed rate at which the blade can cut and the smoothness of the sawn surface. Pitch can be either constant or variable.

RAKE ANGLE - The angle of the face of the tooth measured in respect to a line drawn perpendicular to the cutting direction. Regular- and skip-tooth blades have a 0° rake angle, which gives them a slow, scraping action. A hook-tooth blade has a positive rake angle, which causes it to cut more aggressively.

SET - On blades designed for woodworking, every tooth is set (or bent) left or right, in an alternating sequence, to create a kerf wider than the bladeback. The set of a blade helps prevent binding during cutting. Although carbide teeth are not bent, they are wider than the steel body to which they're brazed. Then they're ground to create a set pattern that helps keep the blade running true.

THICKNESS - The thickness of the steel band measured at the bladeback. (In general, thick blades are wider and stiffer than thin blades.) Thick blades require larger-diameter bandsaw wheels to prevent stress cracks and premature blade breakage.

TEETH - The cutting portions of the blade. Teeth must be sharp, hard, and resistant to both heat and wear. The tip is the sharp part of the tooth that shears away the wood fibers. During sawing, the tooth tip is under tremendous stress and is subject to both heat and wear. The heat produced from friction during sawing can sometimes rise to 400°F on the tip. This occurs because the wood insulates the blade during cutting.

WIDTH - The dimension of a blade from the back of the band to the tip of the tooth. Wider blades are stiffer and resist side-to-side flexing, making them the best choice for resawing. Narrow blades can cut tighter contours.

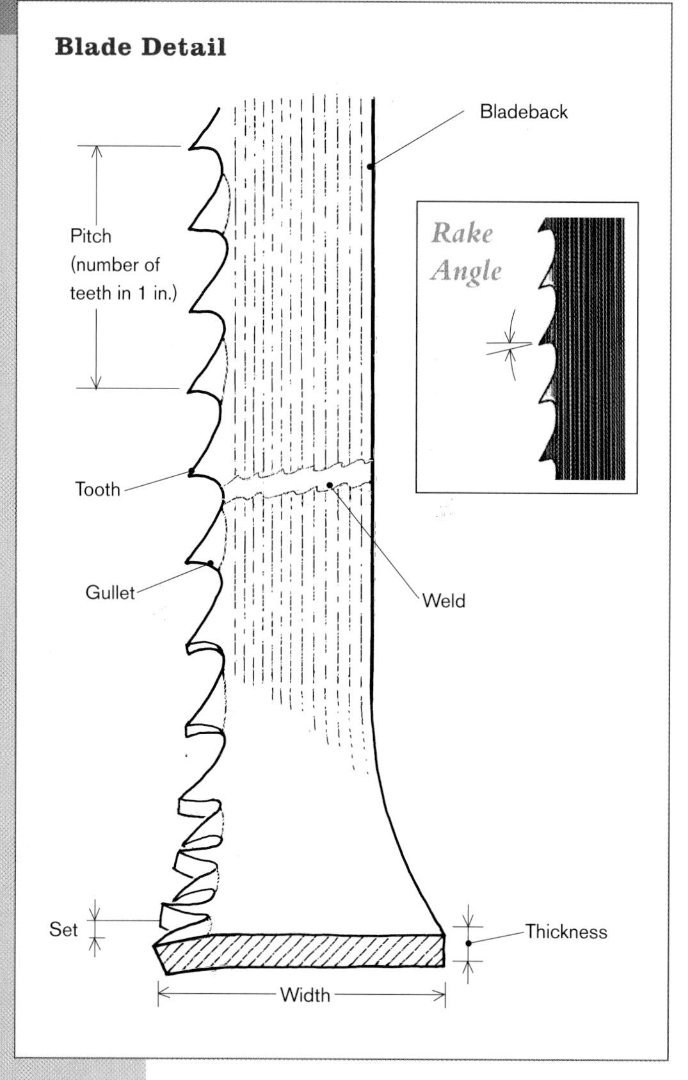

us probably mount a 50-tooth alternate-top bevel (ATB) combo blade on our tablesaw and leave it there until it needs resharpening. That one blade will miter, rip, crosscut, and do just about anything else we need it to do. But this approach doesn't work on the bandsaw, where the blades are much more specialized. The best blade for cutting the contours of a cabriole leg won't accurately resaw veneer. This article will help you develop an arsenal of blades appropriate for the work you do.

Bands of Steel

A bandsaw blade performs a very demanding job. The back of the blade must be soft and pliable to flex around the wheels of the bandsaw at several hundred revolutions each minute, yet the teeth must be hard and resist dulling. Today's blades are stronger, cut smoother, and stay sharp longer than ever before. They also cut with greater efficiency and less feed resistance.

Manufacturers use one of three methods to make the teeth hard and resistant to wear. For carbon-steel and spring-steel blades, teeth are cut into the band, set, and then hardened. In the second method, a band of hard, high-speed steel is welded to a softer band, and the teeth are cut into the harder steel. These are called bimetal blades. For carbide blades, individual carbide teeth are brazed to a flexible steel band. Carbide blades are the most expensive because of the high cost of the material and the process used in making them. Each blade type has advantages and disadvantages, so I'll discuss them individually.

Affordable Carbon-Steel Blades Are Best for Less-Demanding Work Bandsaw blades made entirely of carbon steel are the most common and can be found in almost every consumer woodworking catalog. Carbon-steel blades are sharp, cut well when new, and are available in a variety of widths and tooth forms. They are also inexpensive, which is probably the major reason for their popularity. The main disadvantage of a carbon-steel blade is that it dulls rather quickly, particularly when used for demanding applications such as resawing.

CARBON STEEL

Pros: Inexpensive; weld or braze your own; readily available

Cons: Dulls quickly; cannot be sharpened

Use: Cutting contours in relatively thin stock

Sawing thick hardwood stock places the greatest demands on any blade. If the tooth tip becomes too hot, it becomes soft and quickly loses both its edge and set. Once the set and sharpness are lost, the blade deflects during cutting. The result is that the expensive stock you're sawing is ruined. For these reasons, I use a narrow carbon-steel bandsaw blade only for less-demanding applications such as sawing contours.

Thin Spring-Steel Blades Are Used for Veneer Work Spring steel is most often associated with the cheap, stamped-out blades found on new benchtop bandsaws. Spring steel is soft and flexible, which allow it to bend around the small-diameter wheels of benchtop saws. But because spring steel is so soft, it doesn't hold an edge for very long.

Several years ago, however, a unique spring-steel resaw blade—marketed under the trade name Wood Slicer® and sold by Highland Hardware® (800-241-6748)—was introduced. Instead of being stamped, the teeth on this blade are carefully ground, hardened, and polished. The teeth have a variable spacing that limits harmonic vibration. These blades make smooth cuts, and best of all, the kerf is a mere 1/32 in.—approximately half the kerf width of a typical carbide or carbon-steel blade. This means you'll get more veneer and less waste out of each plank. Additionally, because the Wood Slicer blade is only 0.022-in.-thick spring steel, it easily flexes around the medium-sized wheels of benchtop bandsaws.

STAMPED SPRING STEEL

Pros: Inexpensive; very flexible for use on bandsaws with small-diameter wheels

Cons: Stamped teeth dull very quickly

Use: Light-duty cuts on small bandsaws

Bimetal Blades Offer the Best of Two Worlds The methods used for making bimetal blades are very different from those used for making most carbon-steel and carbide blades. A bimetal blade is actually two steel ribbons that are welded together. The back of a bimetal blade is composed of soft,

BIMETAL
Pros: Cobalt-steel teeth don't readily overheat; high tension means greater beam strength
Con: Don't last as long as carbide
Uses: Demanding applications that generate a lot of heat, such as resawing and cutting thick stock

flexible spring steel; the blade front, where the teeth are milled, is made of much harder high-speed steel. This strip of cobalt steel is welded onto the spring-steel blank before the teeth are cut. When the teeth are cut, all that remains of the cobalt steel is the tooth tip.

This combination produces a relatively inexpensive blade with longer wear than an ordinary carbon-steel blade. Unlike a carbon-steel blade that loses its sharpness and set at 400°F, the cobalt-steel teeth of a bimetal blade can withstand 1,200°F.

Another advantage of a bimetal blade is the beam strength of its spring-steel back, which can withstand great tension. The beam strength (see the top drawing on the facing page) of a bimetal blade, combined with its resistance to heat, has endeared this type of blade to many woodworkers.

Stock Thickness Dictates Blade Pitch

Pitch, the number of teeth per inch (tpi) on a blade, determines the feed rate and the smoothness of the cut surface. A blade with a continuous pattern of teeth has a constant pitch. A blade with teeth that vary in size has a variable pitch.

A blade with a fine pitch has more teeth per inch than a blade with a coarse pitch. A greater number of teeth means that each tooth is small, taking a small bite that leaves a smooth surface. A greater number of teeth also reduces the size of the gullets. Because small gullets can't haul away dust very quickly, a fine-pitch blade cuts slower and tends to get hotter than a coarse-pitch blade.

On a coarse-pitch blade, both the teeth and the gullets are larger, so each tooth bites off a greater amount of wood, and the large gullets can easily remove the sawdust from the kerf.

The major factor to consider when selecting proper tooth pitch is the thickness of the stock. In general, you want a blade that will have no fewer than six and no more than 12 teeth in the stock at any given time (see the drawings above). For example, if you're cutting 1-in.-thick stock, a 6-pitch blade would be a good choice, but a 14-pitch blade would be too fine. However, if the stock were only ½ in. thick, a 14-pitch blade would be best. Although the range of available pitch is broad, from 2 tpi to 32 tpi, wide blades generally have fewer teeth, and narrow blades have a greater number of teeth.

Choosing the correct pitch will substantially increase blade life. Take, for example, a carbon-steel blade, which is easily damaged by overheating. A fine-pitch carbon-steel blade will overheat when used on thick stock because the gullets become packed with sawdust. This causes the blade to dull quickly and lose its set, rendering the blade useless.

Selecting the Appropriate Pitch

You'll get the best cuts when there are between six and 12 teeth in the stock (center). The cut is smooth, and because the sawdust is rapidly carried away, the feed rate can be faster.

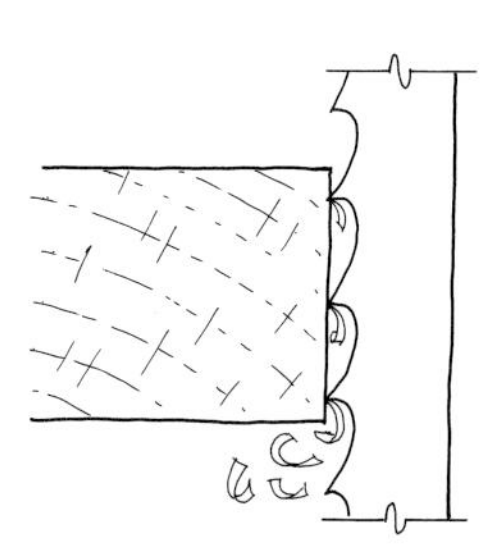
Fewer than six teeth in the stock can cause vibration and a rough cut.

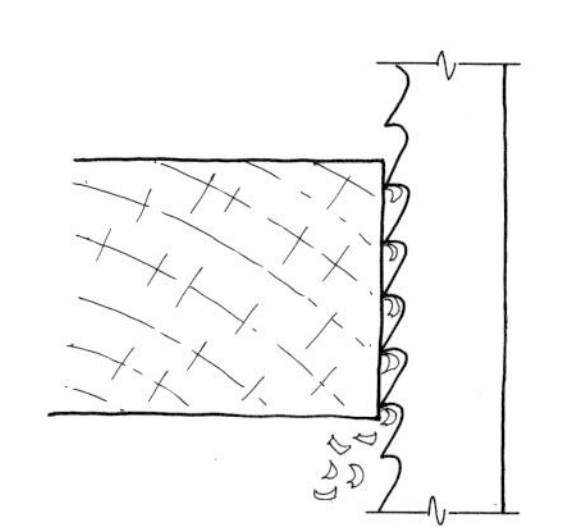
Correct pitch for board thickness results in a fast, smooth cut.

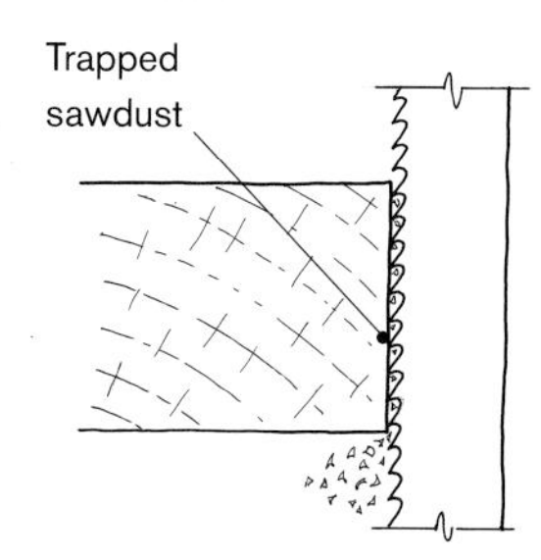

With more than 12 teeth in the stock, the small gullets fill with sawdust, and the blade overheats.

Wider Blades Need More Tension

Beam Strength

A bandsaw blade bows when the beam strength isn't great enough to resist the feed pressure.

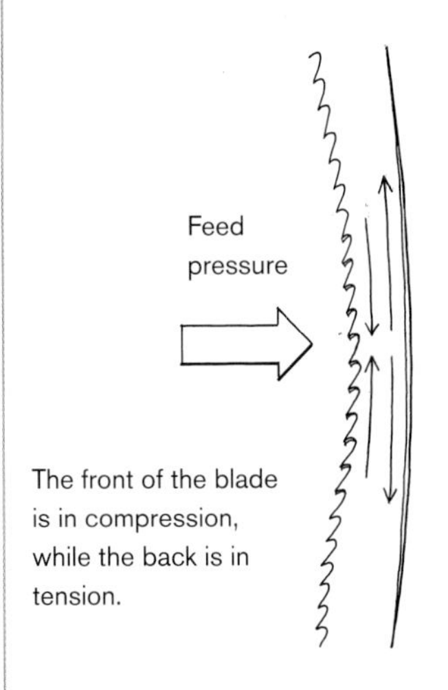

As blades get wider, the steel used for the blades gets thicker. The width of a blade relates to its beam strength–the wider the blade, the stiffer it will be.

A wider blade has more beam strength, but the blade must be properly tensioned. Overtensioning can stress and distort the bandsaw frame, possibly beyond repair. Excessive tension also places potentially damaging forces on the saw's wheels, shafts, and bearings. When resawing, use the widest blade that your bandsaw can properly tension.

Keep in mind that the widest blade a saw can tension may not be the widest blade it can accept. For smaller saws, you'll most likely get better results from the next-size narrower blade.

The most accurate way to determine the proper tension of a blade is to use a tension meter. But a meter has a price tag of around $300, so many choose a simpler route. If you set the upper guides about 6 in. off the table, the blade should deflect under the pressure of a fingertip, but no more than ¼ in. For resawing, the tension should be even a little tighter. Bear in mind that the 14-in. saws common in many small woodworking shops work best with blades no wider than ½ in. Each blade width has a minimum radius that it can cut. Squeezing a blade through a turn that is too tight can break the blade, twist the teeth into the guides (which causes them to lose sharpness and set), or pull the blade off of the saw's wheels, which could damage the teeth or bend the blade. The blade-radius chart below provides the minimum radius that each width of blade can turn. I keep a similar chart posted on my bandsaw.

You may be wondering why you can't mount a narrow blade (such as ¼ in.) and use it for cutting all curves. This does work, but only to a degree. Narrow blades have a tendency to wander. If you try to cut a large radius, such as a 36-in.-dia. tabletop, for example, you'll have a hard time keeping the blade on the line. You'll cut more precisely with a 1-in.-wide blade. However, with practice you'll probably cut a majority of curved work with a ¼-in. or ⅜-in. blade.

How Blade Width Affects the Cutting Radius

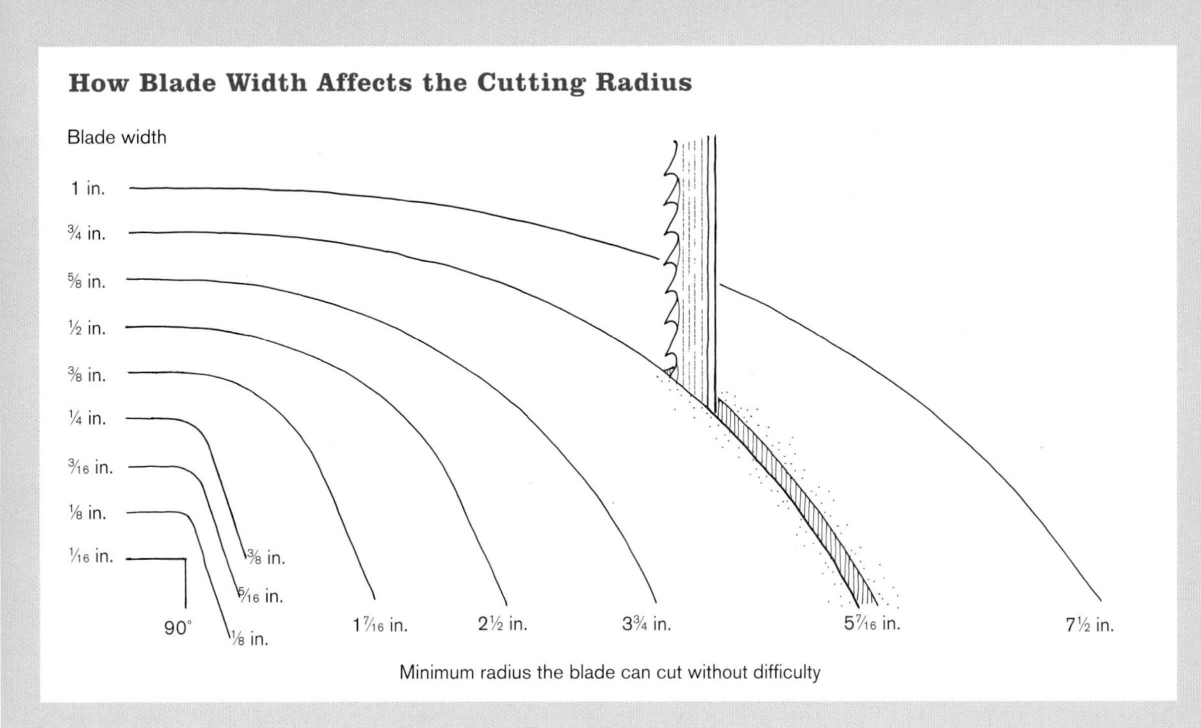

CARBIDE
Pros: Smooth cut; high recommended tension; outlasts carbon-steel blades 25:1
Con: Initial cost is very expensive
Uses: Resawing and other demanding applications

Carbide Blades Are Pricey but Will Last I'm sure that almost every woodworker is familiar with carbide. Carbide cutting tools have almost made high-speed steel a thing of the past. A significant difference between carbide and steel blades is that each carbide tooth is individually brazed onto a strong, flexible spring-steel bladeback. In fact, the recommended tension for a carbide blade is almost twice that of a carbon-steel blade, giving a carbide blade much greater beam strength. The carbide teeth are precisely ground on the face, top, and both sides, which results in truer, more precise cuts.

As you would expect, a carbide bandsaw blade is significantly more expensive than an ordinary carbon-steel blade. However, a carbide blade will typically outlast a carbon-steel blade 25:1, and carbide can be resharpened. Although more expensive initially, a carbide blade is much more economical than a carbon-steel blade, especially for resawing.

Stellite® Is Softer and Less Brittle Than Carbide Stellite is the brand name of a unique type of carbide that is reportedly better suited for woodworking applications. Stellite isn't as hard as regular carbide, but it's also less brittle. This gives Stellite greater shock resistance. Like carbide, Stellite promises longer wear and better-quality cuts.

STELLITE
Pro: More shock resistance than carbide
Cons: Cost; not as hard as carbide
Use: Resawing wide stock

In many other ways, Stellite blades are a lot like carbide blades. The Stellite teeth are brazed onto the band, then precisely

Which Blade Should You Use?

RESAWING POPLAR IS EASY. This 2-pitch bimetal blade makes quick work of softwoods. A carbide blade would also work well but would be more expensive.

SMOOTH OPERATOR. A variable-pitch, hook-tooth carbide blade cleanly slices 1/16-in.-thick veneer from this crotch walnut plank.

Choosing a blade can be confusing until you're familiar with all of the factors. Here are some examples to get you started.

RESAWING 6-IN.-WIDE POPLAR FOR DRAWER PARTS

Option 1: carbide, 3 pitch, hook tooth

Option 2: bimetal, 2 pitch, hook tooth

Comments: Poplar is soft and cuts easily. The bimetal blade would be less expensive, but the carbide blade would last much longer. For greatest beam strength, use the widest blade that your bandsaw can tension.

SLICING 1⁄16-IN.-THICK VENEER FROM A 9-IN.-WIDE CROTCH WALNUT PLANK

Option 1: carbide, 2⁄3 variable pitch, hook tooth

Option 2: spring steel, 3⁄4 variable pitch, hook tooth

Option 3: carbide, 3 pitch, hook tooth

Option 4: bimetal, 3 pitch, hook tooth

Comments: Walnut crotch has dramatic figure and is expensive. I try to get as much veneer as I possibly can from a valuable plank like this. A carbon blade would be my last choice because it dulls quickly. The variable-pitch carbide blade is very expensive, but the cut is incredibly smooth. Both of the carbide blades are stiff and require a strong frame to tension properly. The spring-steel variable-pitch blade is an excellent choice, particularly for saws with wheel diameters less than 18 in. It tensions easily because it's only 0.022 in. thick. This blade cuts incredibly smoothly, and it's relatively inexpensive compared with carbide blades—although you can't expect it to last as long. Best of all, the kerf from this blade is a slim 1⁄32 in., half that of the other blades in this category. You'll definitely get more veneer from this blade.

RIPPING 2-IN.-THICK HARDWOODS

Option 1: carbide, 4 pitch, hook tooth, 1⁄2 in. wide

Option 2: carbon steel, 4 pitch, hook tooth, 1⁄2 in. or 3⁄4 in. wide

Comments: If you have a 14-in. bandsaw, you'll probably get truer cuts with a 1⁄2-in.-wide, 0.025-in.-thick blade than with a 3⁄4-in.-wide, 0.032-in.-thick blade. Your saw stands a better chance of tensioning the thinner and narrower blade.

CUTTING CONTOURS IN 7⁄8-IN.-THICK MAPLE (MINIMUM RADIUS 9⁄16 IN.)

Option 1: carbon steel, 10 pitch, regular tooth, 1⁄4 in. wide

Option 2: carbon steel, 6 pitch, regular tooth, 1⁄4 in. wide

Comments: The 10-pitch blade would create a smoother surface, thus requiring less cleanup of sawmarks.

CUTTING SCROLLS IN 1⁄4-IN.-THICK HARDWOOD (MINIMUM RADIUS 1⁄16 IN.)

Option: bimetal, 24 pitch, regular tooth, 1⁄16 in. wide

Comments: This tiny 1⁄16-in.-wide blade is your only choice for cutting tight contours. You'll need to replace the steel guide blocks with hardwood blocks or Cool Blocks®. This blade can't be used on bandsaws equipped with bearing guides.

THE RIGHT BLADE FOR HARDWOODS. **Ripping hardwoods on the bandsaw is easy with a 1⁄2-in.-wide, 4-pitch blade.**

GOOD FOR MOST CURVES. **A 1⁄4-in., 6-pitch blade can cut most contours, but a 10-pitch blade leaves a smoother surface.**

TIGHT CURVES, CLEAN CUTS. **A 1⁄16-in., 24-pitch blade cuts intricate scrolls with little or no cleanup required.**

ground. And like carbide blades, Stellite blades are expensive.

Different Tooth Forms for Different Jobs

Tooth form refers to the design of the tooth and gullet, specifically the tooth size, shape, and rake, or cutting, angle. The three commonly known tooth forms for cutting wood are regular, skip, and hook. Another form that is gaining in popularity is the variable tooth.

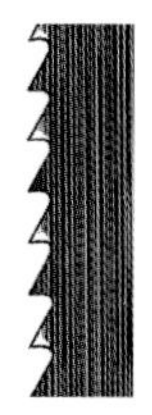

Regular-Tooth Blades The regular-tooth blade, sometimes called the standard form, has evenly spaced teeth for smooth, precise cutting. Teeth and gullets are the same size, and the rake angle is 0°. This combination of features leaves a smooth surface. For cutting curves, a regular-tooth blade is often the best choice because it has the greatest number of teeth. This, combined with a 0° rake angle, gives you a smooth finished surface that requires little cleanup.

The disadvantage of a regular-tooth blade is that the gullets are too small to cut thick stock effectively. Remember that the purpose of the gullets is to haul away the sawdust from the kerf. If you attempt to cut thick stock with a regular-tooth blade, the gullets become full before the teeth exit the stock, which slows cutting and overheats the teeth.

Obviously, a regular-tooth blade is not designed for fast cutting. In fact, if you push the stock too hard in an effort to increase the cutting rate, the cut actually slows down as the gullets become packed with sawdust.

Skip-Tooth Blades As you might assess from the name, the skip form "skips" every other tooth. A skip-tooth blade has fewer teeth and larger gullets than a regular-tooth blade. The large gullets can efficiently carry the sawdust away from the kerf, making a skip-tooth blade fast cutting. Like a regular-tooth blade, a skip-tooth blade also has a 0° rake angle that scrapes the wood away cleanly. But because it has fewer teeth, a skip-tooth blade doesn't cut as smoothly as a regular-tooth blade.

A skip-tooth blade is best suited for resawing and ripping thick stock. It also works well for cutting softwoods. But because the hook-tooth blade is more efficient, the skip-tooth blade is outmoded. Why do manufacturers still produce skip-tooth blades? One sawblade manufacturer said his company still makes skip-tooth blades mainly because—short of sending people a free hook-tooth blade—it's difficult to convince people to change.

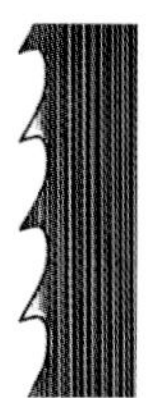

Hook-Tooth Blades The hook tooth is really a further development of the skip tooth. A hook-tooth blade has large gullets and teeth like that of a skip-tooth blade, but the teeth have a positive rake angle that makes them cut more aggressively. Because of this aggressive nature, a hook-tooth blade has less feed resistance than a skip-tooth blade. It is a great choice for resawing and ripping thick stock. A hook-tooth blade is my choice for general resawing, such as sawing thick planks into thin drawer parts. The coarser pitch and positive rake angle of a hook-tooth blade make quick work of any hardwood.

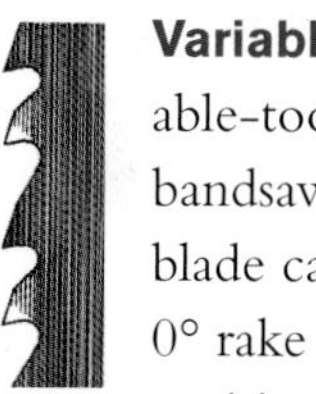

Variable-Tooth Blades The variable-tooth blade is a hybrid among bandsaw blades. A variable-tooth blade can have regular teeth with a 0° rake angle or a more aggressive, positive rake angle. But the unique feature of this type of blade is that the tooth size and spacing vary on the same blade. This means that both the teeth and gullets vary in size but not in shape. The unique design dramatically reduces vibration; the result is a quieter blade and a very smooth cut.

To understand how this works, it's helpful to think of a bandsaw blade as a string on a musical instrument. A bandsaw blade is under tension, just like the strings on a violin but for different reasons. You want a string on an instrument to vibrate so that it produces a sound. This is called harmonic vibration. But you want to limit vibration on a bandsaw blade because vibrations create a rough surface on the stock. By varying the tooth and gullet size, you effectively limit the vibrations and create a smoother surface.

When sawing veneer from a plank of valuable hardwood, a hook-tooth blade will do a great job, but a variable-tooth blade will leave a much smoother finish.

Tooth form affects the performance of the blade more than any other factor. A regular-tooth blade gives the smoothest cut; a hook-tooth blade cuts aggressively; and a variable-tooth blade cuts both smoothly and aggressively.

The Right Blade Choice

Rather than thumbing through the pages of an industrial bandsaw blade catalog, it's much easier to narrow the blade choices based upon the types of cuts you'll be making. For every job, it's important to consider the blade width, pitch, and tooth form. I always begin by selecting the blade width. Width is determined by the type of cut you're making—whether you're sawing a straight line or a curve. Tooth pitch is dictated by the thickness of the stock you'll be cutting, and tooth form influences how aggressively or smoothly the blade will cut.

To get the most out of your bandsaw, you'll have to change blades often from wide to narrow or from few teeth to many. Each type of blade is best for a certain kind of cutting. You must decide which is more important to you—speed or smoothness. You can't get the best of both in the same blade. However, you can select a blade that is a good compromise.

LONNIE BIRD is a frequent contributor to *Fine Woodworking* magazine and teaches woodworking in Tennessee.

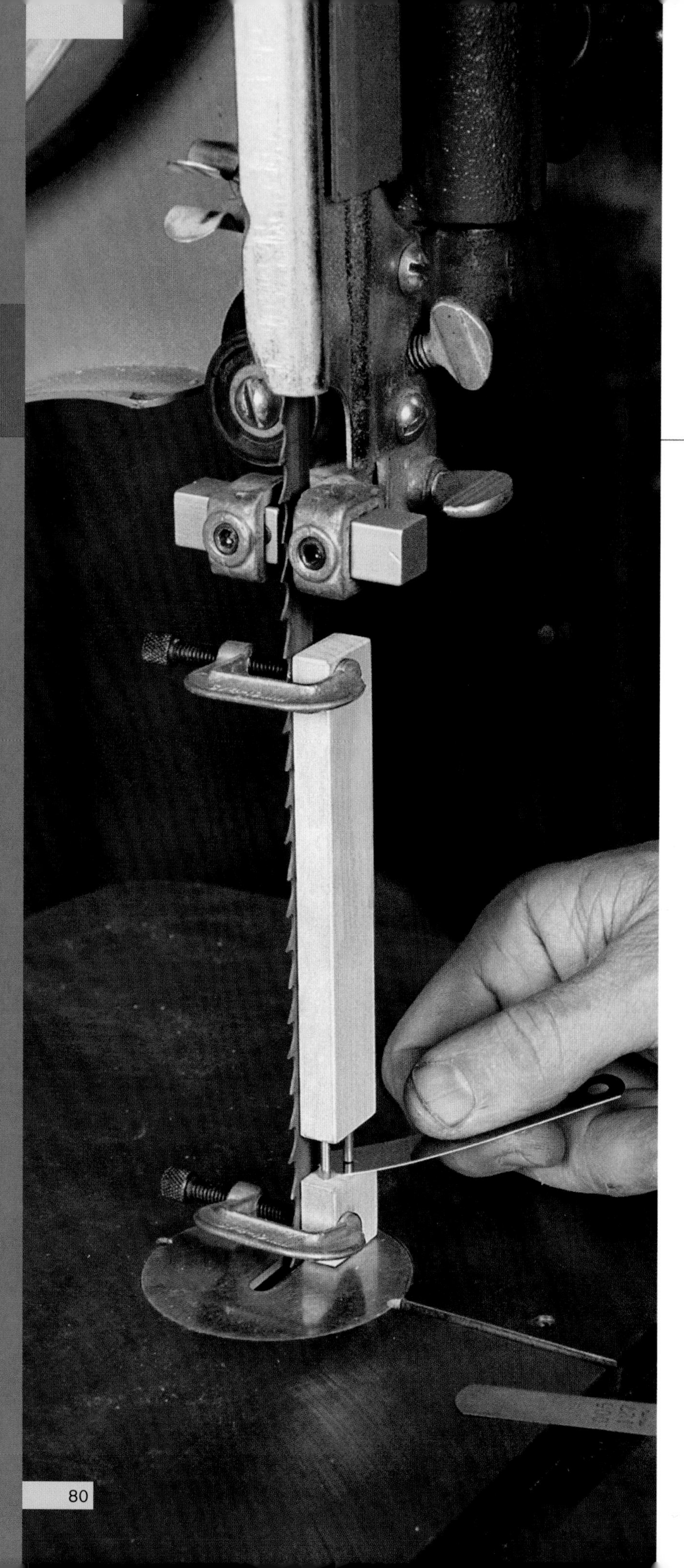

Shopmade Tension Gauge

BY JOHN WHITE

A bandsaw blade that's not properly tensioned is going to cause problems no matter how well the rest of the saw is tuned up. And if you're doing an especially tough job, like resawing a wide board or ripping thick hardwood stock, the problems are going to be even worse. In many shops, however, the only way to measure blade tension is to use the gauge built into the back of the bandsaw. Yet these gauges are notoriously inaccurate. That usually results in a blade that is undertensioned.

A blade under too little tension leads to all sorts of grief. It can bow backwards and sideways, causing the blade to cut slowly. It easily wanders from a cut line. And when resawing, the cut often takes on an unwelcomed barrel shape.

But too much tension on a blade creates its own set of headaches. It can overstress the wheels and bearings of the bandsaw, and sometimes the frame, too.

That's where this unassuming little tool comes in handy. Used with an ordinary automotive feeler gauge, it allows you to set the correct tension on your steel blade quickly. And it works with any bandsaw or blade.

Making the Gauge

There's nothing fussy about making the tool. A fine-grain hardwood is best here because you want the ends to have hard, flat surfaces. Maple, birch, and beech are all good choices.

Start by cutting the hardwood to a ⅜-in.-thick by ¾-in.-wide by 5⅜-in.-long strip. Then, in one end of this strip, drill a 3/32-in.-dia. hole for the alignment pin, making the hole 2 in. deep. A T-shaped fixture I made came in handy here, allowing me to clamp the strip so that it is perpendicular to the table.

Next, using a fine-toothed crosscut blade on the tablesaw, carefully cut a ¾-in.-long block from the drilled end of the strip. Before cutting this short block, make index marks on both sides of the cut line. These marks enable you to line up the two pieces in their original orientation when the gauge is assembled. Polish the cut face of the short block by running it across very fine sandpaper laid on a flat surface. Do not round the face. For accuracy, it must be absolutely flat.

Next, drill a ½-in.-deep, 3/32-in.-dia. hole for the measuring pin in the end of the long block, next to the hole for the alignment pin. Make the measuring pin by cutting a 6d finish nail ⅞ in. long. It's a good idea to round the working end of the pin slightly, shaping it with a file and then polishing it with a fine stone or emery paper. Then tap the pin into its hole, making sure you don't accidentally use the alignment-pin hole.

Now cut the alignment pin. Once again use a 6d finish nail, but this time cut it 1⅝ in. long. This pin should fit tightly in the short block but slide in the hole in the long block. The simplest way to achieve this is to place the pin in a drill chuck with about ¾ in. exposed and then slightly reduce the pin diameter with a fine file as it spins. The mild steel in the nail will cut quickly, so check your fit frequently and stop as soon as you have a sliding fit. The gauge will be more accurate if there is no excess play. Smooth the ends of the pin and tap its larger-diameter end into the short block, making sure the index marks are facing each other.

That's all there is to making the gauge. But before it can be used, you need to know just how much tension to apply to the blade. And as I learned, the answer to

Bandsaw Tension Gauge

Not much bigger than a pen, this tension gauge can be made for pennies with a small piece of hardwood and two finish nails.

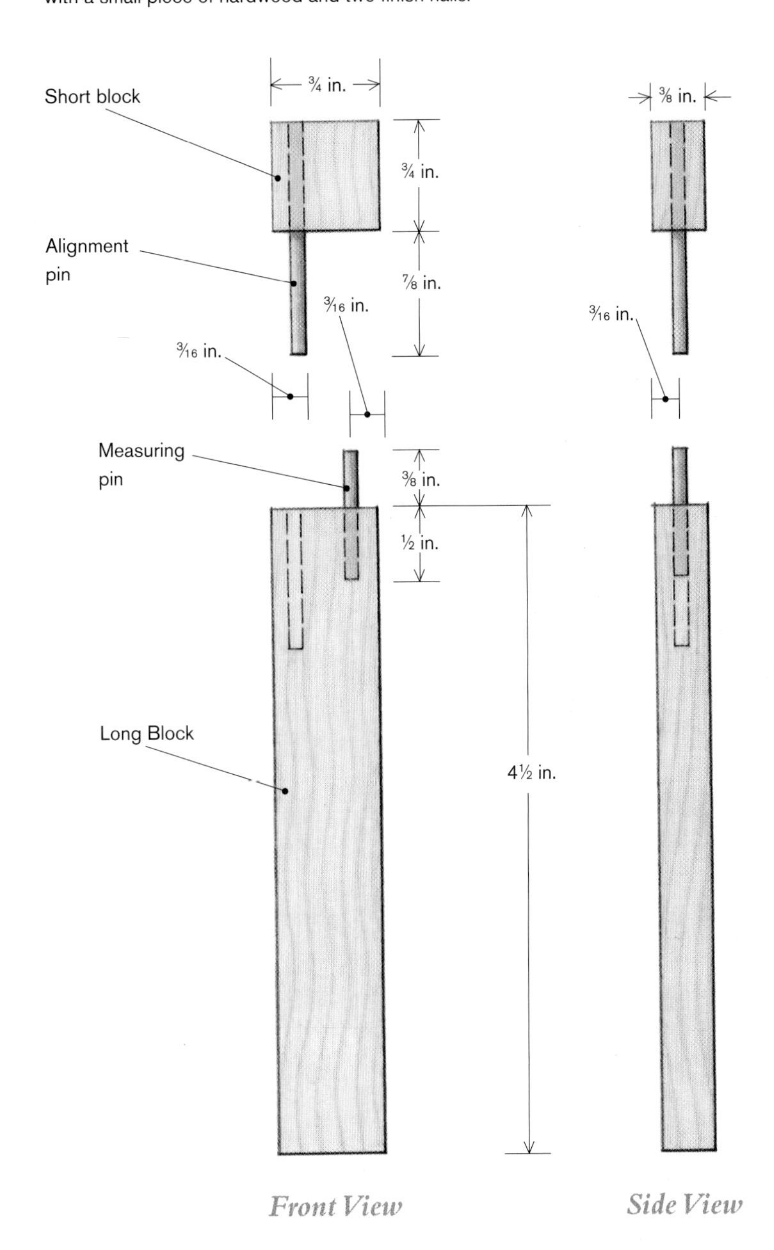

that question depends on the kind of bandsawing you're doing.

How Much Tension?

For tough jobs like resawing or cutting thick stock, blade makers suggest a maximum blade tension of 15,000 lbs. per square inch (psi) when using a consumer-grade bandsaw (such as the 14-in. Delta or its cousins). Use the same tension regardless of whether the blade is carbon steel, bimetal, or carbide-tipped.

Keep in mind that you don't have to use such high tensions for all work. On a consumer-grade saw, you'll extend the life of the blade, tires and bearings if you lower the tension to about 8,000 psi when cutting thin stock and softwoods.

Any tension beyond 15,000 psi could cause problems with the structure of a consumer-grade saw. But for some of the heavy-duty, industrial-quality bandsaws, the blade can be tensioned up to 30,000 psi if it's going to be used for resawing.

It's interesting how "psi" translates into the number of pounds of tension applied to the blade and the wheel. A ½-in.-wide by 0.025-in.-thick blade requires a pull of about 188 lbs. to achieve a tension of 15,000 psi. And because both the cutting and returning halves of the blade are under tension, the bandsaw's tensioning spring must push up the top wheel with twice this force, a total of 375 lbs.

The suggested tension, measured in psi, remains the same no matter what size blade you're using. So, for example, because the ⅜-in.-wide blade has a smaller cross-sectional area than the ½-in.-wide blade, you need only about 140 lbs. of pull (280 lbs. on the top wheel) to produce 15,000 psi of tension.

But knowing how much to tension the blade is only half the story. You also have to be able to tell how much tension is actually being applied, which is where this gauge comes in.

The Gauge Is Easy to Use

The best way to determine blade tension is to measure the amount the blade stretches as it's pulled taut. As you might expect, steel doesn't stretch easily. In fact, a 5-in. length of blade stretches only 0.001 in (that's one thousandth of an inch) for every 6,000 psi of tension that's applied.

Based on this principle, several companies make a tension meter with a dial indicator

DRILL THE ALIGNMENT-PIN HOLE. **Cut a hardwood strip to size, then drill a hole in the end for the alignment pin. A T-shaped fixture keeps the strip square to the table.**

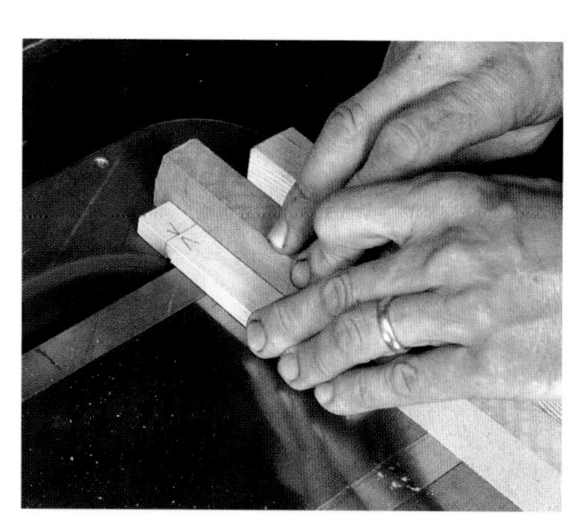

MARK AND CUT. **After adding a pair of index marks, the tablesaw is used to cut off the short-block portion of the tool.**

FILE DOWN THE ALIGNMENT PIN. **If the alignment pin (a 6d nail) is to slip smoothly into a mating hole, it must be filed down a bit as it spins in the drill press.**

Using the Gauge

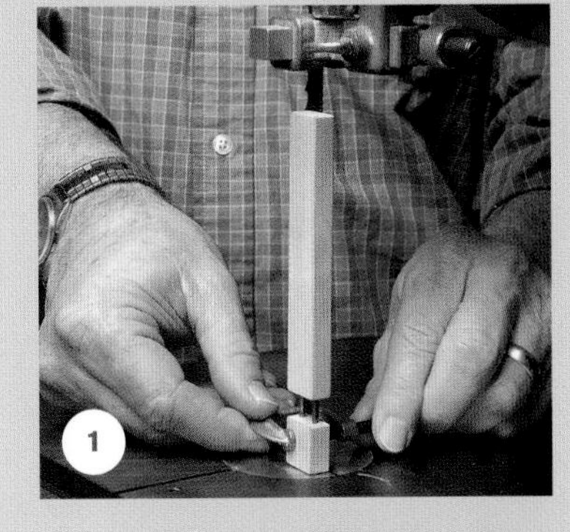

RELAX THE BLADE TENSION. Then clamp the short block of the gauge to the side of the blade, just behind the teeth.

LEAVE ROOM FOR THE FEELER GAUGE. Clamp the long block to the blade so that a 0.005-in. feeler gauge fits between the block and the measuring pin.

ADD TENSION TO THE BLADE. Once a 0.008-in. feeler gauge slips between the short block and the measuring pin, back off a bit.

that reads the amount of blade stretch. However, at $130 to $320, these instruments don't come cheap.

That's the beauty of the gauge I designed. It measures stretch, just as the expensive version does, but you won't have to stretch your budget to get one. To check the accuracy of my gauge, I set it up in tandem with a top-of-the-line tension meter. The measurements on my gauge were within 10% to 15% of the readings on the tension meter.

And using the gauge is as simple as making it. First install a blade on your bandsaw, then tension it lightly and adjust the tracking and guides while moving the blade by hand. When the blade is running properly, power up the saw and let it run for a minute or two to warm up the blade and tires, then cut the power and unplug the saw.

Next, back off the tension until there is just enough pull on the blade to prevent it from going slack and slipping off the tires. Usually this is going to be a little below the tension mark for a ⅛-in.-wide blade on the machine's built-in scale.

Raise the upper blade guide out of the way and lightly clamp the gauge to the side of the blade just behind the teeth (see photo 1 above). The measuring pin should be toward the back of the blade. Loosen the upper clamp and pinch a 0.005-in. feeler gauge between the measuring pin and the long block (see photo 2 above). Leave the tool in place, and tighten both clamps.

Now recheck the gap with the feeler gauge. It may change slightly from the torque of the clamps, but it isn't important that it be exactly 0.005 in. A starting gap of 0.004 in. or 0.006 in. will work just as well.

A Spring with Spunk

If you tighten the tension gauge on any small, consumer-grade bandsaw, such as the 14-in. Delta, there's a good chance you won't come close to reaching the 15,000 psi of tension that's recommended for resawing on these lighter-weight machines. With a ½-in.-wide blade, you're likely to find that running the saw's indicator off the end of the tensioning scale, beyond the ¾-in.-wide blade setting, increases the gap by only 0.001 in. (6,000 psi) or perhaps not at all.

If you continue to crank down on the tensioning knob, the gap (and the tension) is going to finally and suddenly increase but only because you've crushed the coils of the spring until they're touching.

A BETTER SPRING. The spring on a consumer-grade bandsaw (top) won't be able to apply as much tension as the aftermarket spring made by Iturra (bottom).

However, running the saw with the spring collapsed will damage the saw. That's because the spring also serves an important secondary function as a shock absorber.

If your saw can't reach 15,000 psi of tension, it's because the springs on these smaller machines go soft quickly, and a fatigued spring exerts far less force than it was originally designed to apply, no matter how far it is compressed.

The answer is to buy a new spring. Iturra Design in Jacksonville, Florida (888-722-7078), makes one from a better grade of steel, and there's more of it, so it lasts longer. Plus it is stiffer, so you can add more tension. For $14.95 the Iturra spring is a good investment.

By the way, you can extend the useful life of any spring (and blade) if you remove most of the tension when the saw isn't being used. This is especially important with the Iturra spring. Just be sure to remember to retension the blade before turning on the saw.

To tension the blade, begin by choosing a feeler gauge that equals the width of the starting gap, plus an additional 0.001 in. for each 6,000 psi of tension you want to apply to the blade. For example, if your starting gap is 0.005 in. and you want 15,000 psi of tension, start with a 0.008-in. feeler gauge (see photo 3 on p. 83). This is going to give you 18,000 psi of tension, but don't worry; it's going to be adjusted lower almost immediately.

With the 0.008-in. feeler gauge in hand, increase the blade tension while using the gauge to check the gap under the pin. When you reach 18,000 lbs., the feeler gauge is going to fit just under the pin. Once there, you can back off a little on the saw's tension adjustment to end up in the range of 15,000 psi.

Now you can mark your saw's tensioning scale at the pointer, noting the width of the blade. The recalibrated scale will allow you to tension the same-width blade quickly in the future without using the gauge every time. To guard against the spring becoming weaker, it's a good idea to use the tension gauge and feeler gauges to recheck the scale occasionally.

With a little practice, this little bandsaw tool will allow you to set the blade tension in less than five minutes. And with tension set just right, you can look forward to getting better performance from your bandsaw.

JOHN WHITE is a woodworker and machinist who maintains the *Fine Woodworking* workshop.

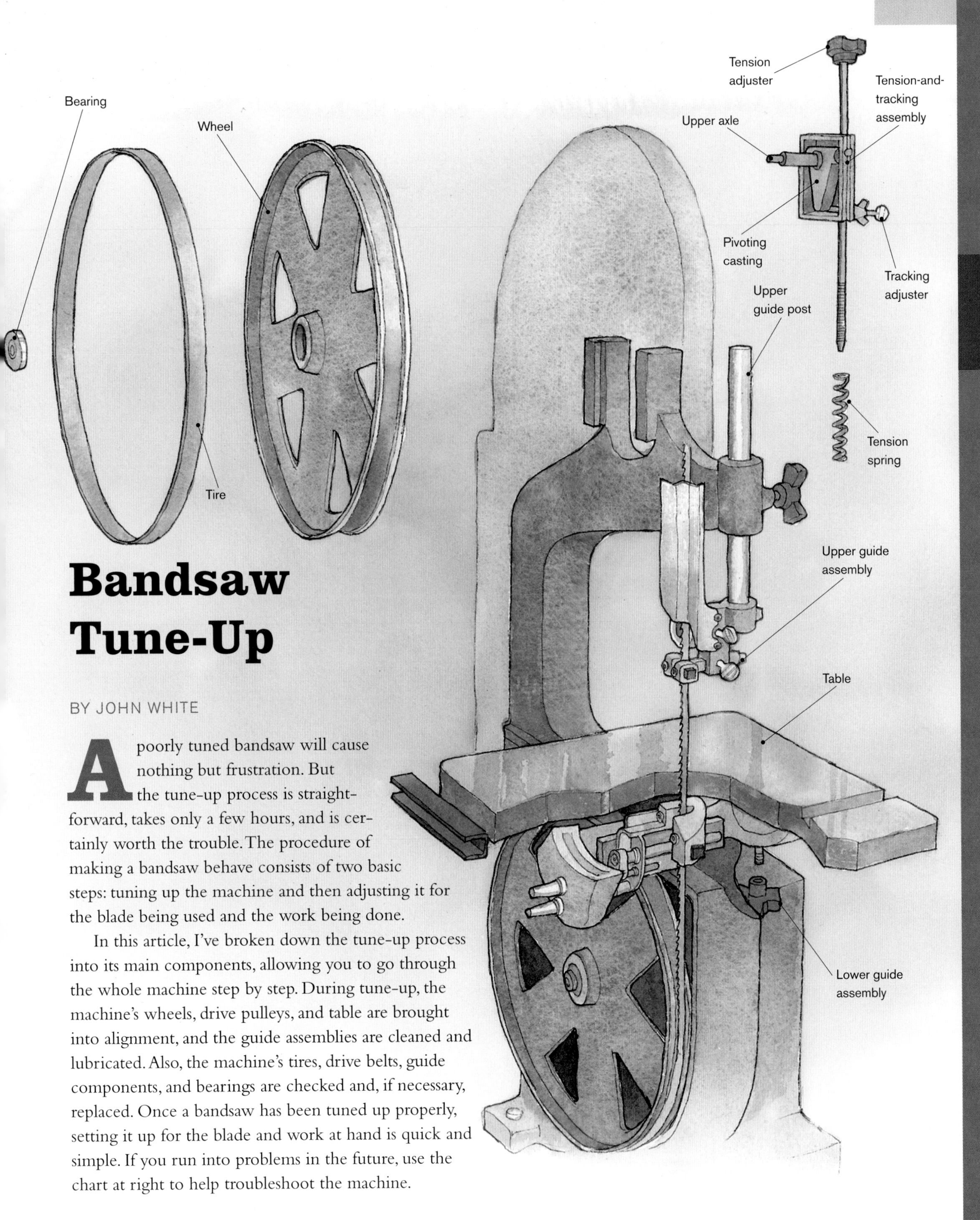

Bandsaw Tune-Up

BY JOHN WHITE

A poorly tuned bandsaw will cause nothing but frustration. But the tune-up process is straightforward, takes only a few hours, and is certainly worth the trouble. The procedure of making a bandsaw behave consists of two basic steps: tuning up the machine and then adjusting it for the blade being used and the work being done.

In this article, I've broken down the tune-up process into its main components, allowing you to go through the whole machine step by step. During tune-up, the machine's wheels, drive pulleys, and table are brought into alignment, and the guide assemblies are cleaned and lubricated. Also, the machine's tires, drive belts, guide components, and bearings are checked and, if necessary, replaced. Once a bandsaw has been tuned up properly, setting it up for the blade and work at hand is quick and simple. If you run into problems in the future, use the chart at right to help troubleshoot the machine.

JOHN WHITE is a woodworker and machinist who maintains the *Fine Woodworking* workshop.

Sources

Iturra Design
888-722-7078
for a free catalog

TROUBLESHOOTING GUIDE

Symptom	Possible Cause		See
Vibration at high speed	• Pulleys on motor are worn or bent • Drive belt is stiff or worn	• Wheel-bearing failure • Thrust-bearing failure	Wheels, Blade guides
Vibration at low speed	• Wheels are bent or misaligned • Dust buildup on tires • Tires are cracked or worn	• Tire is lifting off wheel • Blade is cracked or kinked or has a bad weld	Wheels, Tires
Blade doesn't stay centered on wheels	• Tires are grooved, hardened, or worn • Wheel-bearing failure	• Wheels are misaligned • Tracking mechanism is slipping or bent	Tension, Wheels, Tires
Blade doesn't cut straight	• Blade is dull • Fence is not aligned for drift	• Worn guide blocks • Low blade tension • Poorly adjusted guides	Tension, Blade guides
Cut is barrel-shaped	• Blade is dull or too narrow • Feed rate is too fast	• Low blade tension • Poorly adjusted guides	Tension, Blade guides

Tension

The tensioning-and-tracking assembly controls the position of the upper wheel. Remove the upper wheel and cover to get at the assembly. Inspect the pivoting casting that supports the axle for cracks or bends from overtensioning. A cracked or bent casting should be replaced. If the axle is loose, which is common, don't worry—it will tighten up when tension is applied. Use light oil to lubricate the pivot pin.

After a few years of use, the slides on both sides of the square main casting will probably have a step worn into them—use a file to smooth down the worn faces and the sharp edges left on the top of the grooves in the casting. Use a stick lubricant on the slides before sliding the assembly back into the frame.

The original tensioning spring on a 14-in. bandsaw is almost always crushed, making it impossible to tension the blade properly. The spring can be replaced with a heavy-duty version from Iturra Designs (see sources) without having to remove the upper wheel and blade cover.

The last step in servicing the top end of the saw is to remove the tension and tracking bolts—clean the threads with a wire brush and round off the ends with a file. Use a stick lubricant on the bolts before you reinstall them.

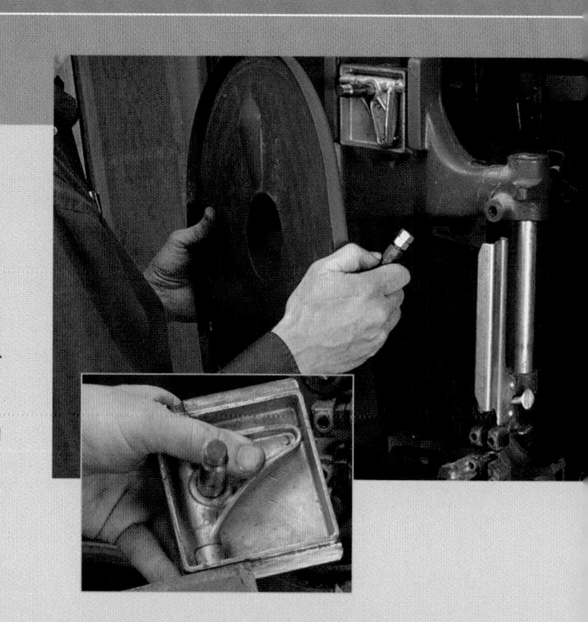

REMOVE COVER AND CLEAN UP TENSIONER. With the cover removed, it is easy to access the tensioner. Once the tensioner is removed, check to be sure the axle isn't bent, and then file the slides on both sides smooth.

Wheels

Saving the upper wheel aligned directly above the lower wheel allows the bandsaw blade to track better and puts less stress on the saw and the blade. On a 14-in. bandsaw, checking the alignment is easy. Remove the table and lay a long straightedge across the faces of both wheels. If the wheels are out of alignment, you'll see a gap between the straightedge and one wheel. On a Delta saw, the wheel alignment is adjusted by adding or removing shims on the upper axle. On Jet® and most other Taiwanese-made saws, the upper wheel can't be shimmed without placing excess pressure on the wheel bearings. These saws are aligned by shimming behind the lower wheel. Iturra Designs sells inexpensive sets of graduated shims for both Delta and imported bandsaws.

CHECK THE ALIGNMENT. **Begin by placing a long straightedge (White uses a 4-ft. level) across both wheels. Then adjust the tracking mechanism to bring the upper wheel parallel to the straightedge.**

SHIM OUT THE UPPER WHEEL. **If the rims of both wheels aren't touching the straightedge, use shims to bring them into alignment. On Delta saws you can shim the upper wheel; for Taiwanese-made saws, shim the lower wheel.**

CHECKING AND REPLACING WHEEL BEARINGS

To test the wheel bearings, remove the saw's blade and rotate the wheel through several revolutions with the tip of a finger against one of the spokes. You may feel a slight drag, but the motion should be smooth and silent. Even small amounts of roughness or a grinding sound indicate a contaminated bearing. If there is only a small amount of catching, the saw is still usable, but new bearings should be installed soon. If there is continuous roughness or grinding noises, the saw shouldn't be used until new bearings have been installed.

OUT WITH THE OLD, IN WITH THE NEW. **A bandsaw wheel has two bearings: Even if only one is failing, they should both be replaced. The wheel bearings must be tapped out with a hammer and punch (left). When installing a new bearing, gently tap it into place using a soft hammer against the outer race (right).**

Tires

Tires are simply oversized rubber bands. But they should be checked regularly, because the rubber becomes worn, cracked, or hardened and can cause tracking problems and vibration. A tire should have an obvious crown and be smooth and free of grooves. Press your thumbnail into the tire; it should press in easily, and the surface should spring back. A lack of springback is a sign that the tire has hardened and needs to be replaced. To remove a tire, use a screwdriver to lift it over the rim. If the old tire was glued on, clean off the adhesive using acetone. The new tire should snap into a groove in the rim of the wheel.

INSTALLING A TIRE. Stretch the tire over the wheel (top). To even out the tension on the tire, slide a screwdriver between the wheel and tire, then rotate the wheel while holding the screwdriver in one spot (above).

Table

To get square cuts on a bandsaw, the table must be aligned square to both the sides and back of the blade. To align the table, first back off the blade guides and then adjust the blade to full tension and proper tracking. Place a square on the side of the blade and adjust the stop bolt (see the photo below) to square up the table. Once the table has been adjusted, zero out the pointer on the table-tilt scale. To square the table to the back of the blade, loosen the table bolts from underneath, remove the table, and place shims between the trunnions and the table casting. This process may take a little trial and error, but you only have to do it once, and it is definitely worth the time.

Aligning Front to Back

SQUARE THE TABLETOP TO THE SIDE OF THE BLADE. Tip the table and adjust the stop bolt mounted on the trunnion support casting.

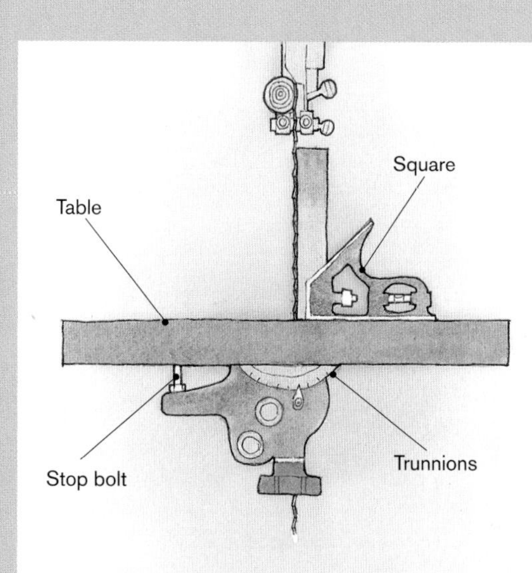

Aligning Side to Side

THERE'S ONLY ONE WAY TO SQUARE THE TABLE TO THE BACK OF THE BLADE. You need to insert shims between the table and the trunnions until the blade and the table are aligned. While you have the top off, be sure to clean up and lubricate the trunnions.

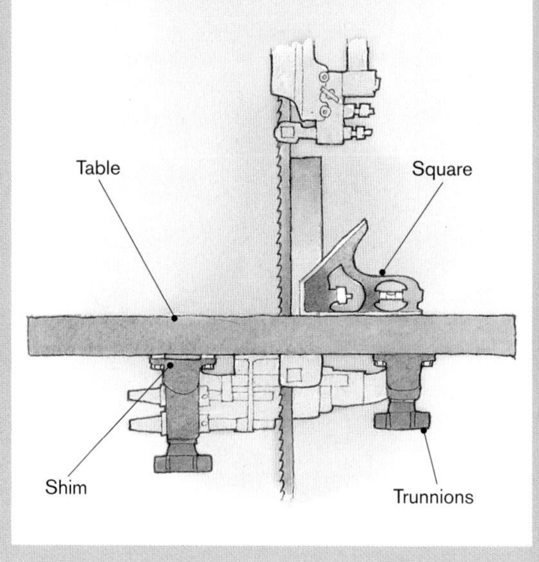

Blade Guides

ALIGN THE UPPER THRUST BEARING. **The blade should ride along only the outer edge of the bearing. If one face is scarred from use, flip over the bearing and use the back face.**

The guide assemblies on a 14-in. saw are mechanically simple but have a number of parts that can wear out or jam up. Start by replacing any thrust bearing that is noisy or won't rotate freely. Then remove the bearing support and guide-block holders, file off any paint and burrs, and inspect all parts for cracks or worn threads. Remove all of the setscrews and round off their ends with a file—the smoothed ends will hold better. Remove the knob that locks the guide post and shape the tip of its threaded end to match the groove in the guide post. Clean and lubricate the threads and the other parts of the guides as you reassemble them.

The guide blocks should be smooth, flat, and square. Clamp the blocks in the holder with their faces touching; there should be no gaps between the blocks.

The lower guide assembly on the Delta 14-in. bandsaw is more complex than the upper guide assembly, but the same logic applies to tune-up. The lower guide assembly on a Taiwanese-made saw is tuned up the same way as the upper guide assembly.

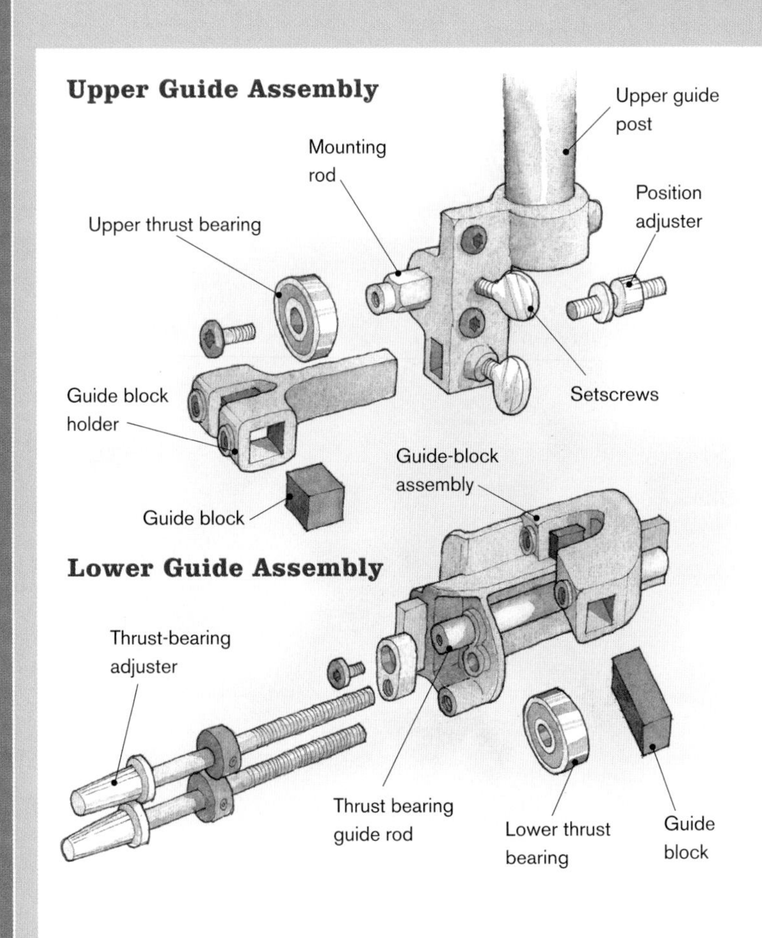

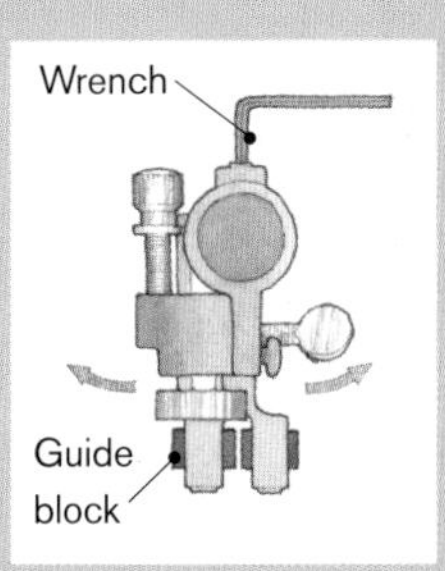

ALIGN THE GUIDE ASSEMBLY WITH THE BLADE. **Loosen the Allen screw on the upper guide assembly and adjust the assembly until the faces of the guide blocks are parallel to the blade.**

REPLACE THE LOWER THRUST BEARING. **Remove the nut on the end of the guide rod and slide off the tube to free up the bearing. Clean up everything and, if necessary, slide a new bearing into place.**

Installing a Bandsaw Blade

A bandsaw is not properly tuned until you have installed the blade and made sure it is tensioned and tracking properly and that the guides are set correctly. Following the steps here makes this a quick and straightforward process.

Position the Blade

With the saw unplugged, pull back the guides and the thrust bearings and place the new blade on the wheels. Raise the upper guide assembly to clear the stock you'll be cutting by ¼ in. to ½ in.

Tension and Track

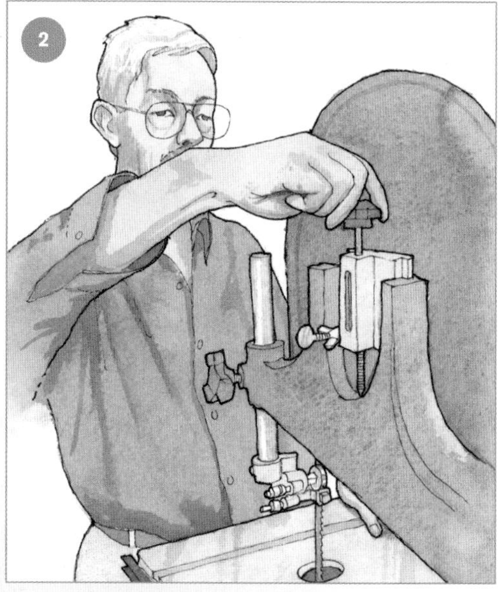

Rotate the upper wheel by hand while alternately increasing the tension and adjusting the tracking to keep the blade centered on the upper wheel.

Turning the tracking adjustment in adjusts the blade toward the back of the wheel.

Adjust the Guide Assemblies

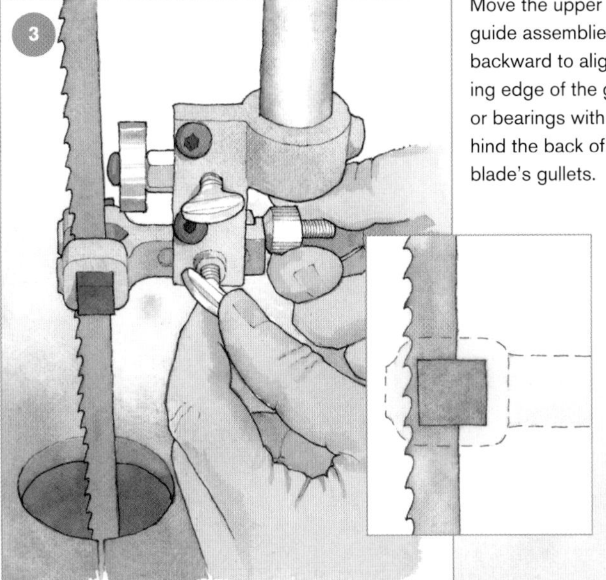

Move the upper and lower guide assemblies forward or backward to align the leading edge of the guide blocks or bearings with or just behind the back of the sawblade's gullets.

Adjust the Guide Blocks

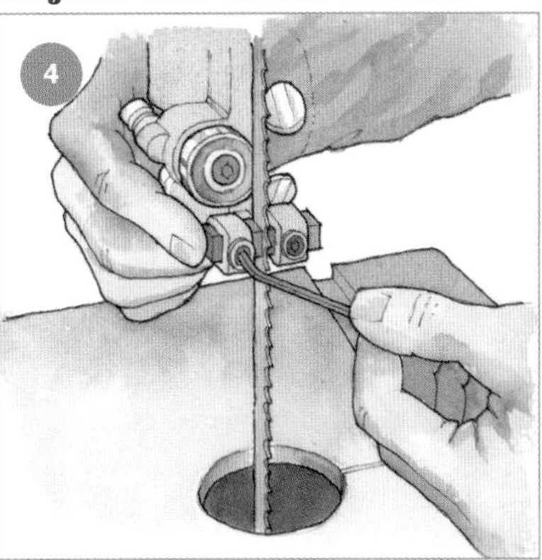

Move one of the guide blocks or bearings in each assembly so that it just touches the side of the blade. Lock it in place. Double-check that the block or bearing doesn't reach beyond the back of the blade's gullets. Bring the second block of each assembly against the blade. A soft block can be locked in place touching the blade. Hard blocks or ball-bearing guides should be spaced away from the blade with a single piece of paper. Rotate the blade by hand to check that a bad weld or kink in the blade won't cause problems.

Position the Thrust Bearing

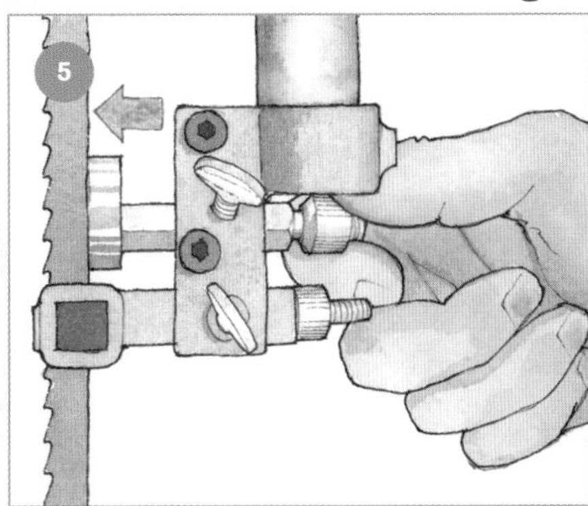

Bring the upper and lower thrust bearings forward to just barely touch the back of the blade. Rotate the blade by hand to make sure everything turns smoothly.

Align the Fence for Drift

Once the blade is following the line, hold the stock in place and turn off the saw. Use a marker to draw a line on the tabletop along the edge of the stock. Reinstall the fence and adjust its angle parallel with the mark on the table.

About your safety: Woodworking is inherently dangerous. Using power tools improperly can lead to injury. Follow all of the instructions set forth by the manufacturer. Don't try to perform these operations until you are certain that they are safe for you and your machine.

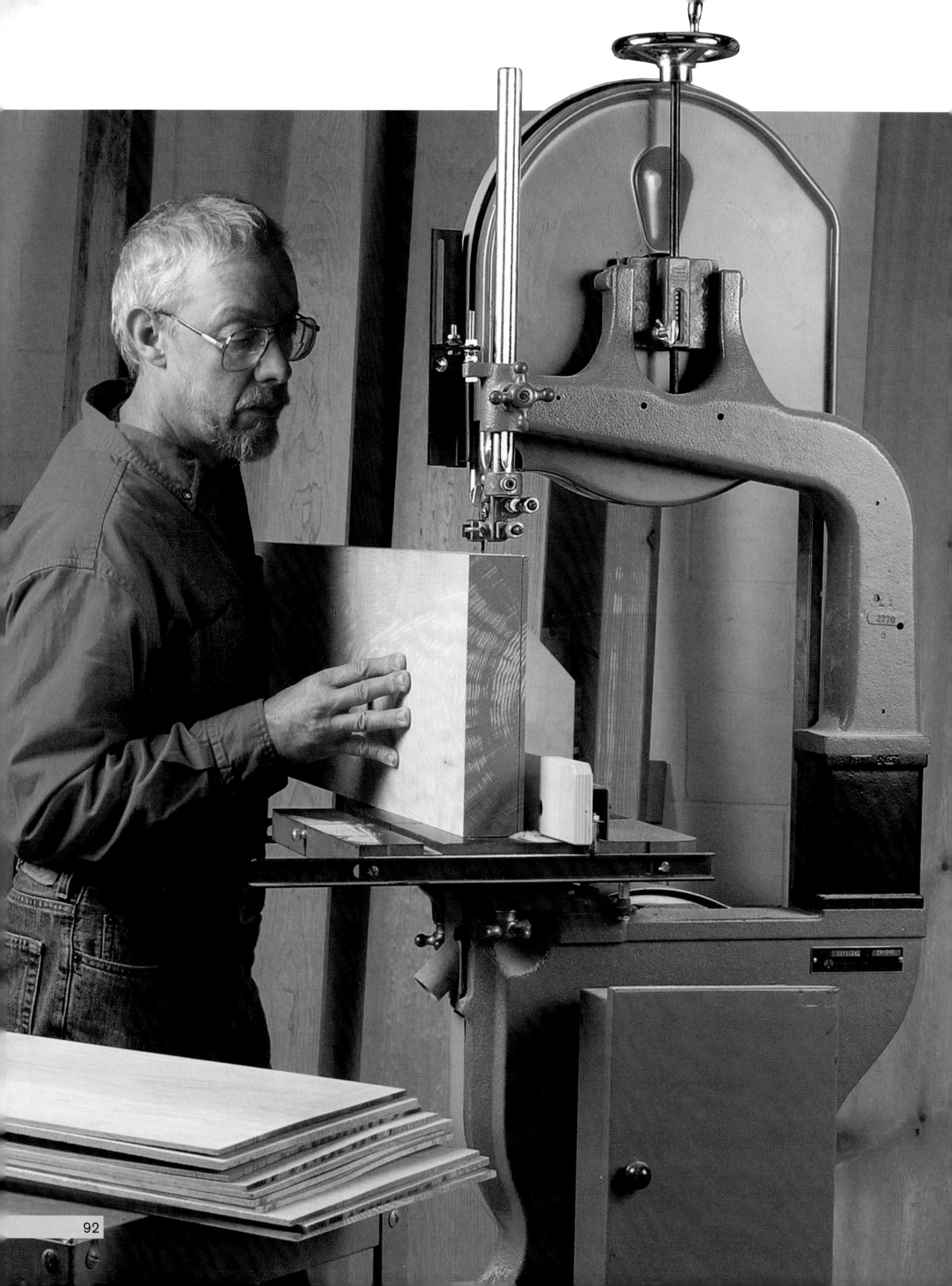

Soup Up Your 14-in. Bandsaw

BY JOHN WHITE

The classic 14-in. cast-iron bandsaw, developed by Delta Machinery back in the 1930s, was designed mostly for making curved pattern cuts in relatively thin boards. Delta still makes that saw here in the United States. Over the years, it has changed little. Along the way, it even served as a model for several of the Asian-made clones currently sold in the United States, including a couple now marketed by Delta.

Nowadays, though, it seems more woodworkers are looking to push the limits of 14-in. cast-iron bandsaws by using them to resaw wide boards. And many are finding out that it's not always easy to do. The feed

A Riser Block Increases Cutting Capacity

Most 14-in. cast-iron bandsaws can be outfitted with a riser-block kit that's available as an option. Adding a riser block increases the resaw capacity of the machine from about 6 in. to 12 in. With this increased capacity, though, you'll also need longer blades.

THE RISER BLOCK MOUNTS between the upper and lower frames of the saw. Two pins help position the block on the frame. All three parts are held together with a heavy-duty bolt, washer, and nut.

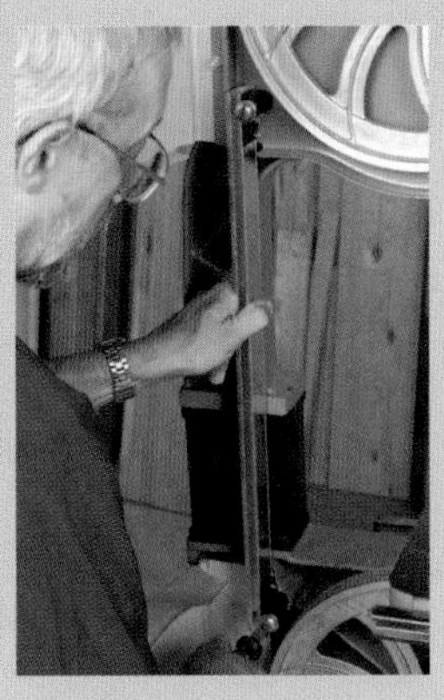

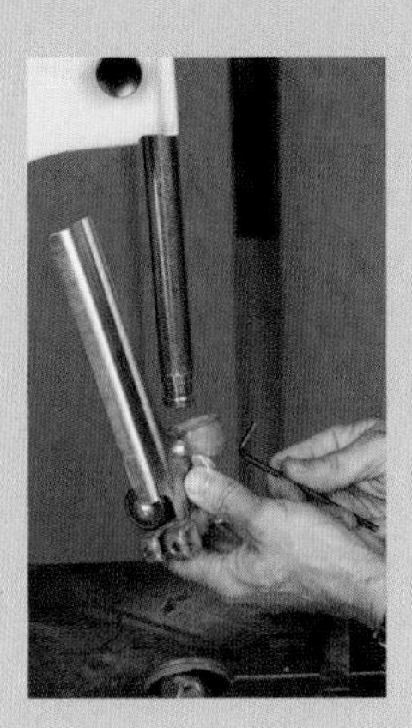

ATTACH THE REPLACEMENT BLADE GUARDS and the guidepost. The riser-block kit includes longer blade guards for both the back and front of the saw and a longer guidepost.

The Right Blade for Resawing

Resawing can be next to impossible with the wrong blade. To haul away all of the sawdust created while resawing, you need a blade with big gullets, which means fewer teeth per inch. Also, hook teeth are generally better for resawing.

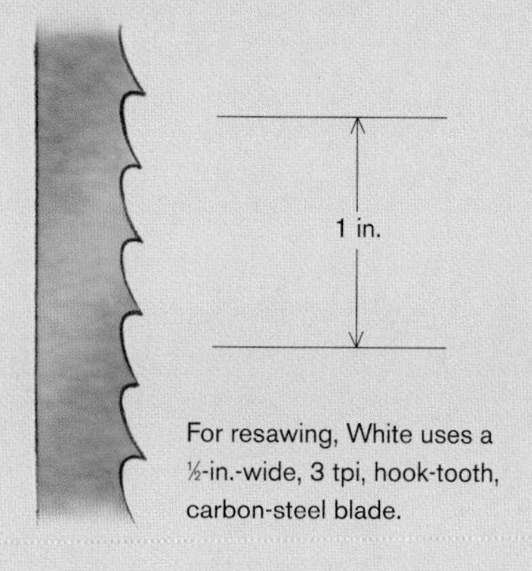

For resawing, White uses a ½-in.-wide, 3 tpi, hook-tooth, carbon-steel blade.

REPLACE A WORN-OUT SPRING. If your cut wanders, chances are that the spring is not providing enough tension to the blade.

rate is annoyingly slow, the motor often bogs down, the blade can drift off line, and the cuts sometimes end up far from square.

But don't trade in the saw yet. I've found that by making some relatively minor modifications, a typical ¾-hp to 1-hp, 14-in. cast-iron bandsaw can be transformed into an effective resawing machine. Indeed, my upgraded machine resaws 12-in.-wide maple boards with little effort.

The Basic Upgrade

The basic upgrade adds a riser block to increase the resaw capacity, a resaw blade, a heavier tension spring, and a tall fence. This upgrade will enable you to start resawing stock as wide as 12 in.

Riser Block Doubles the Capacity Most manufacturers of 14-in. cast-iron bandsaws offer a riser-block kit as an optional accessory that increases the resaw capacity from roughly 6 in. to about 12 in. The kit also includes a longer guidepost and a pair of longer blade guards. The block is bolted between the upper and lower frames. All of the kits include extralong bolts to account for the added length.

A Good Resaw Blade Is a Must Perhaps more than anything else, a good-quality resaw blade can go a long way toward improving the resawing capabilities of a bandsaw. Resaw blades have large gullets that carry away the considerable sawdust that's generated when cutting through wide boards. Avoid blades with small gullets and lots of teeth because they aren't designed to cut wide stock.

In general, a ½-in.-wide hook-tooth blade (2 tpi or 3 tpi) will work fine. I've also had good experience with both the Timberwolf™ and Wood Slicer blades.

THE FACE OF THE FENCE EXTENDS just past the sawblade teeth. That way, should the cutoff piece curl outward, it can't push the board away from the fence during the cut.

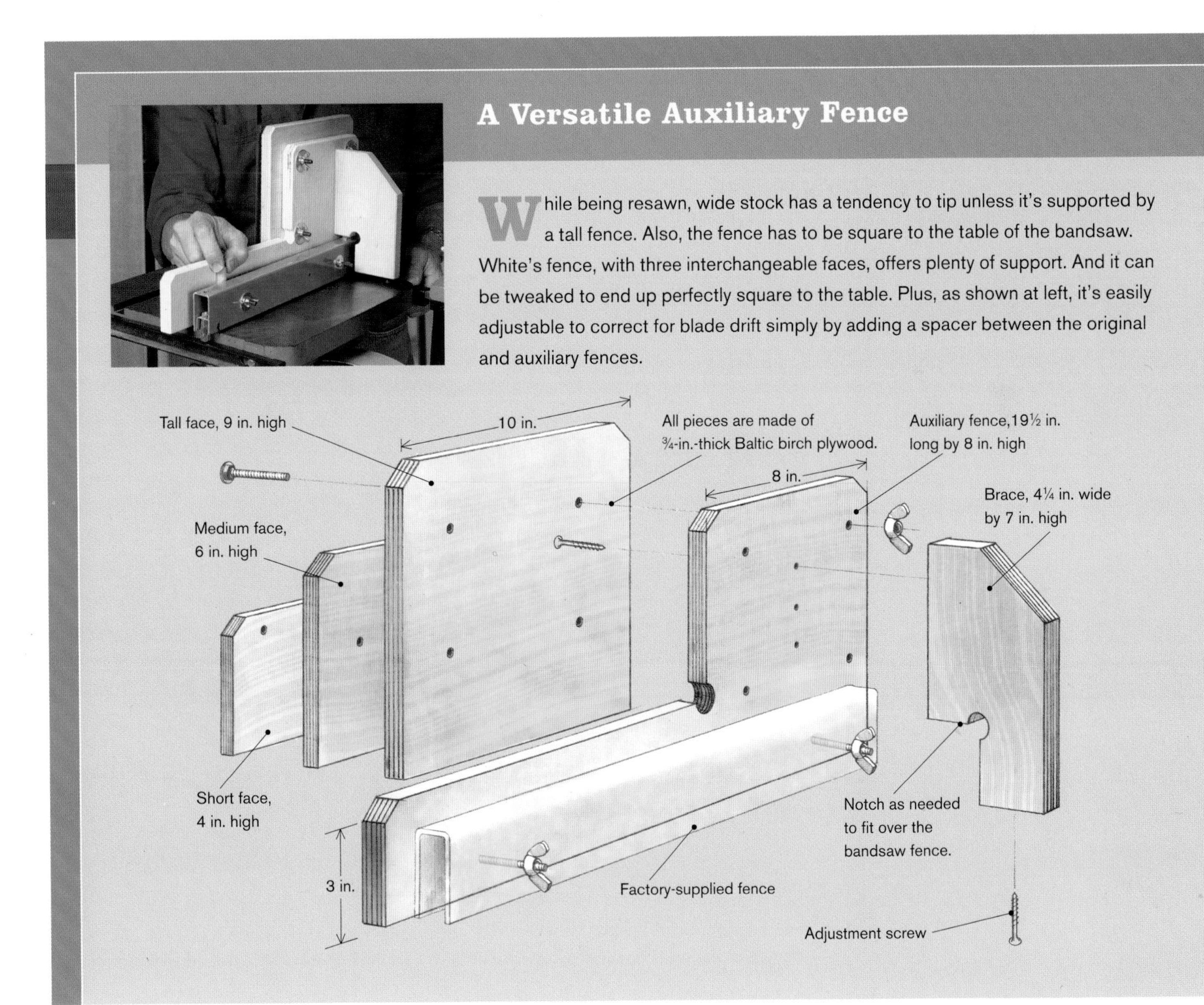

A Versatile Auxiliary Fence

While being resawn, wide stock has a tendency to tip unless it's supported by a tall fence. Also, the fence has to be square to the table of the bandsaw. White's fence, with three interchangeable faces, offers plenty of support. And it can be tweaked to end up perfectly square to the table. Plus, as shown at left, it's easily adjustable to correct for blade drift simply by adding a spacer between the original and auxiliary fences.

Sources

RESAW BLADES
Highland Hardware
Wood Slicer
800-241-6748

Suffolk™ Machinery
Timberwolf
800-234-7297

Woodworker's Supply
Carbon steel, hook-tooth, 3 tpi
800-645-9292

HEAVY-DUTY TENSION SPRING
Iturra Designs
888-722-7078

WEAK TRUNNION SUPPORTS EQUAL TILTING TABLE. **Pushing down on the saw tables with the heel of his hand, White discovered that cast-aluminum trunnion supports were easily deflected out of square.**

Shopmade Trunnion-Support Plate Eliminates Table Flex

Delta's 14-in. cast-iron bandsaw has a trunnion support made from cast iron. But most 14-in. cast-iron saws have a cast-aluminum trunnion support. Aluminum castings are usually thinner than the Delta version and tend to flex easily, so anything other than a lightweight board can make the saw table tip a bit. That's not helpful when you want a square cut.

Adding a simple plywood plate to a saw with a cast-aluminum trunnion support will eliminate almost all of the table movement. On the downside, adding the plate sacrifices ¾ in. of resaw capacity, but I feel the trade-off is worth it. The plate is a rectangular piece of ¾-in.-thick birch plywood, with a notch to clear the blade path. It also has a slot to catch an eyebolt that attaches to the back edge of the table. Tightened with a wing nut, the eyebolt locks the table against the 90° stop bolt, virtually guaranteeing that the table stays square to the blade.

The plate dimensions on the facing page should work on most 14-in. cast-iron saws. If you need to revise the dimensions, design the plate so that it ends up 2 in. wider than the trunnion support and projects ½ in. past the back edge of the table.

To locate the holes for the two pins that stick up from the cast-iron saw frame, slide the plate down over the bolts that hold the plate in place and strike the top of the plate with a mallet. Remove the plate and use the dents left by the pins to center and drill the pin holes.

To fully support the trunnion casting, you may have to slip a washer between the plywood and the casting at the two outboard ends of the casting where the bolts that lock the trunnions come through.

To account for the thickness of the plywood, use longer bolts to hold the casting and the blade-guide assembly.

The eyebolt that holds the saw table tightly against the stop bolt slips through a hole drilled in the lip that reinforces the edge of the table. A sharp drill will easily go through the thin cast iron. I cut away a quarter of the loop in the eyebolt to make it easy to fit through the hole. A spacer on the bolt allows the wing nut to clear a flange cast into the top edge of the saw's lower frame.

INSTALLING THE SUPPORT

A LOOK UNDER THE HOOD. **Before the trunnion support can be removed, the table of the bandsaw must be removed.**

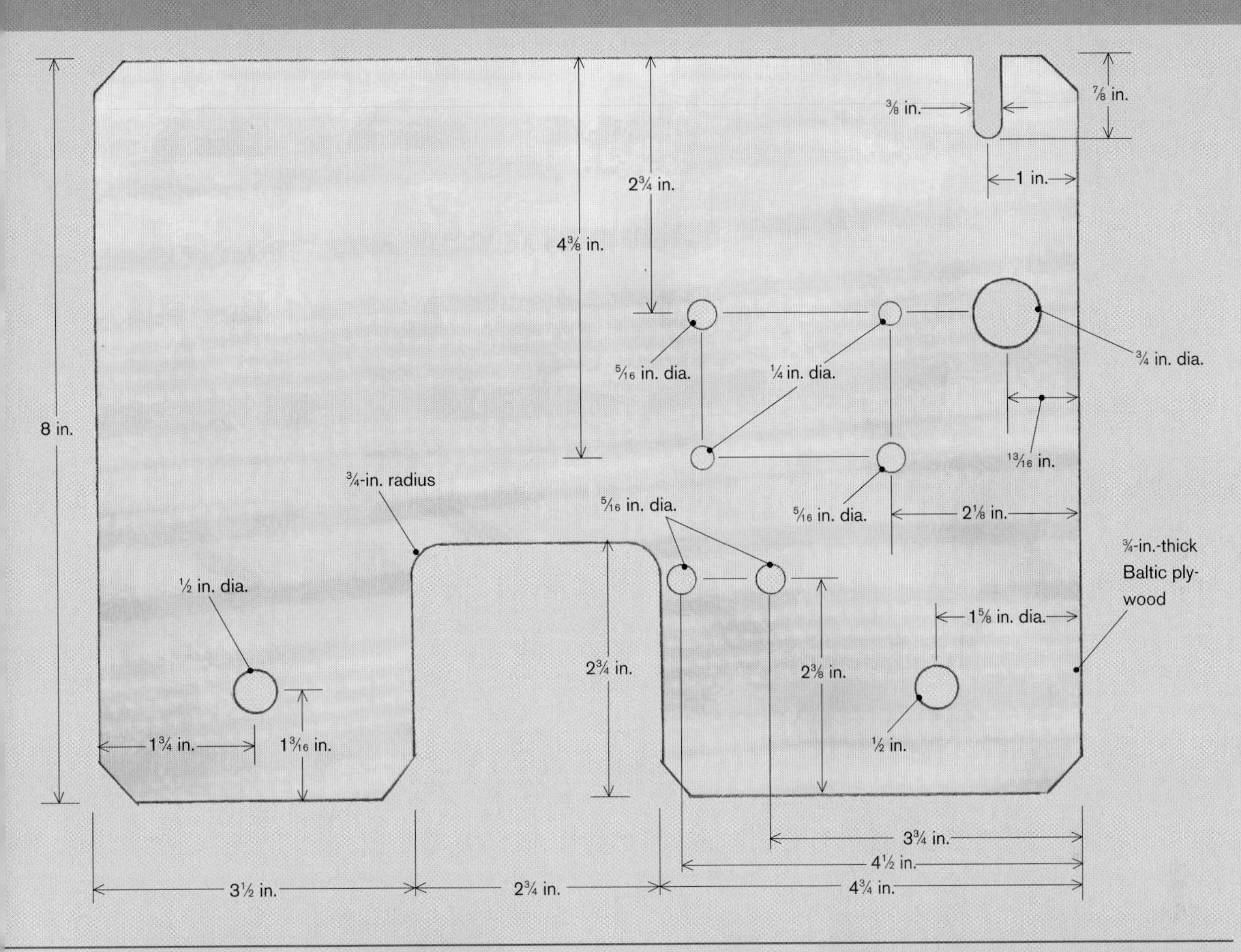

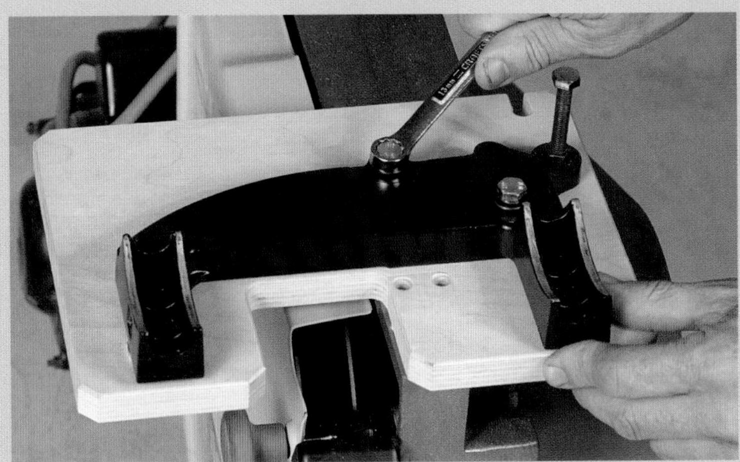

TRUNNION SUPPORT ADDS STRENGTH. A single piece of plywood is all it takes to beef up the aluminum trunnion support. One drawback is that the resaw capacity of the saw is reduced by ¾ in.

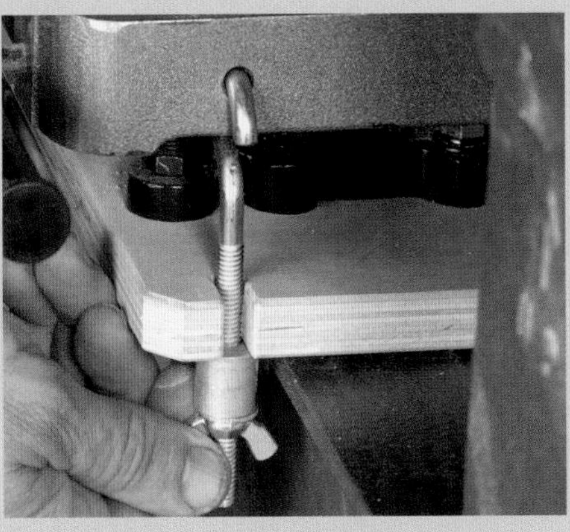

EYEBOLT KEEPS THE TABLE PARTS SNUG. A hole bored in the edge of the saw table accepts an eyebolt that slips into a slot in the plywood. Tightening the wing nut keeps the underside of the table firmly against a single adjustable stop bolt.

Dedicated Resawer Edition

Anyone doing a lot of resawing should consider adding infeed and outfeed supports along with a lower base. The supports help keep the board from tipping off the saw table, while the lower base places the board at a more comfortable height for resawing and has storage for blades.

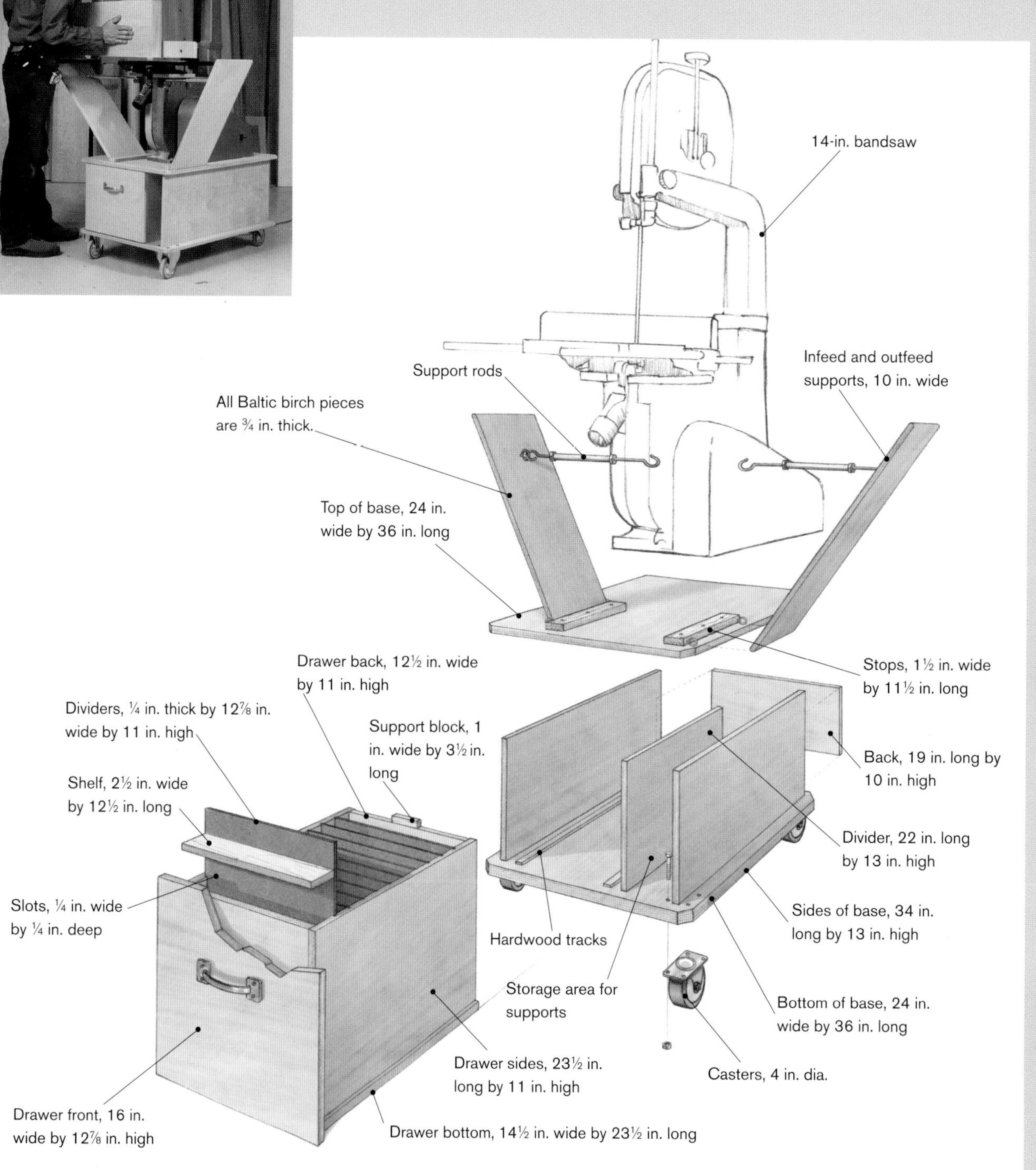

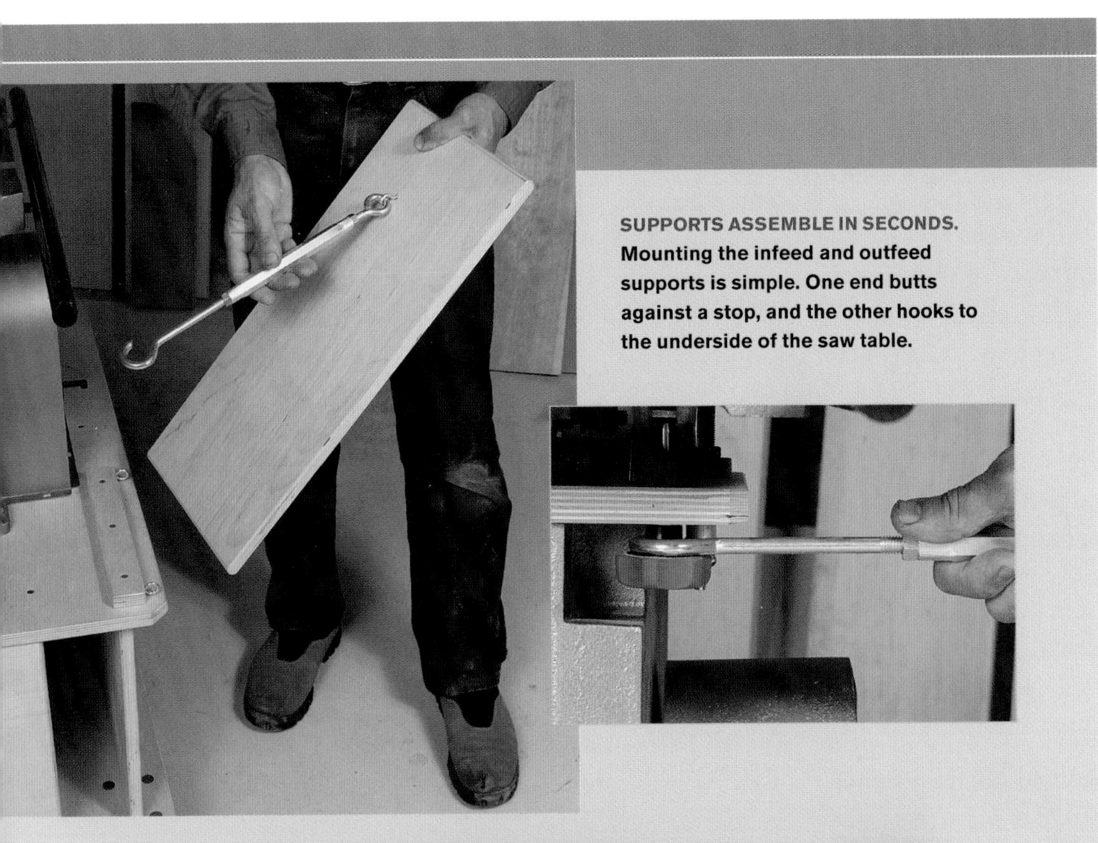

SUPPORTS ASSEMBLE IN SECONDS. Mounting the infeed and outfeed supports is simple. One end butts against a stop, and the other hooks to the underside of the saw table.

DRAWER OFFERS STORAGE. The single drawer can hold several coiled bandsaw blades, while the shelf in front is a handy place for small parts. The supports slide into the compartment on the right.

Beefier Spring Adds More Tension To provide the best possible cut, a bandsaw blade must be tensioned properly. A blade lacking adequate tension is more likely to wander from the cut or produce a bowed cut when resawing.

The source of the blade tension is a compressed spring behind the upper wheel. It's not uncommon for this spring to lack some vitality. When that's the case, the spring can't apply enough tension.

The solution is to replace the original spring with one that has more muscle (see the bottom photo on p. 94). But first check with the supplier to make sure your saw can handle the additional load; otherwise, you might end up with a burned tension mechanism or prematurely worn bearings.

Add an Auxiliary Fence I also made a tall auxiliary fence to support a board during resawing (see the drawing and photo on p. 95). The fence is short in length but sturdy enough that it remains square to the table when pressure is applied. In addition, it is easy to adjust the fence angle to eliminate blade drift—the propensity for a bandsaw blade to wander from a straight line during a cut. One thing to note: If the factory-supplied fence on your bandsaw is of little value, you'll need to get a better fence before you can add this resaw fence.

My auxiliary fence extends to just past the trailing edge of the blade. There's a reason for this short length. Thick stock often has a fair amount of tension in the wood, even when carefully dried. When you're resawing, the tension in the wood is released, sometimes causing the offcut to bend or twist into the fence, which means you'll have to push pretty hard to keep the board against the fence. By using a fence that is short in length—much like that on a European-style tablesaw—you can keep the uncut portion of the board firmly against the fence while the offcut is free to bend or twist into the open air.

The fence I installed has two main components: a back piece with a brace that bolts to the original fence and a set of interchangeable faces, each one of a different height. The idea here is to use a face that is narrower than the board to be resawn but wide enough to support the board adequately during the cut. That way, both for safety and maximum blade support, the blade guard along with the upper blade guide can be lowered close to the top edge of the board.

With this fence, it's easy to adjust for blade drift by adding spacers between the original fence and the auxiliary fence, changing the angle of the fence relative to the blade. Dowels or small hardwood blocks work fine as spacers. On some blades, the drift can be considerable. One blade I use needs a ¾-in.-thick spacer before it cuts straight.

For accurate cuts, the auxiliary fence should be square to the table. Sometimes, however, a factory-supplied fence won't be quite as square as you'd like. To correct for this out of squareness, mount thin shims (I use strips of aluminum flashing attached with double-faced tape) to the back of the fence. The shims should be long enough to bear on any spacer that is added to correct for blade drift. Once the auxiliary fence has been squared to the table, adjust the screw on the bottom of the brace to give the fence added support.

Advanced Upgrade

I recommend the advanced upgrade for a bandsaw being used exclusively for resawing. It adds infeed and outfeed supports to make it easier to support wide and long boards. Except for the ones on the Delta models, the trunnion support is going to need some beefing up (see the sidebar on pp. 96–97). Also, the factory-supplied base is replaced with a shorter, shopmade base that lowers the saw table to a more comfortable resawing height.

Add the Infeed and Outfeed Supports

The infeed and outfeed supports are simply ¾-in.-thick plywood panels that tip out from the top of the base. The panels are held in place by support rods made out of large turnbuckles. The rods attach to the saw by slipping around the shank of the knobs that lock the table trunnions. The turnbuckles make it easy to fine-tune the supports to the bandsaw's table height. The panels rest against stop blocks on the saw's base. Eyebolts screwed into the stop blocks help position the piece on the base. And the support rods just hook over the knobs. Removing the panels takes only a few seconds.

To enlarge the stock steel base enough to accept the infeed and outfeed supports, add a piece of plywood between the steel base and the cast-iron frame (or make a larger base like the one shown on pp. 98–99).

To make a support rod, replace the right-hand threaded eyebolt on each turnbuckle with a longer bolt that has an eye large enough to slip around the shank of the trunnion lock-knob. Once the eyebolts have been adjusted, add nuts to lock them in place. Because one of the threads in each turnbuckle is left-handed, you'll need a left-handed nut to lock that side. I had no problem finding all of the hardware I needed at the local hardware store.

The infeed support should tilt at about a 45° angle. That way, the top end of the support can't be easily pushed and lifted by the board it's holding up. The outfeed support, however, should be installed at a more upright angle so that you can move the outfeed table's stop block away from the saw and clear the cover for the drive belt. The top end of the support cannot lift because, on the outfeed end, the drag of the board pulls against the support rod.

There's a simple way to determine the length of the supports. Make them longer than necessary. Hold them in position against a straightedge placed across the saw table, mark, and cut. The top and bottom edges of the supports are rounded over with a ⅜-in.-radius router bit.

Make the Base For ripping and resawing, the table of a bandsaw should be close to the height of a tablesaw, not the 42-in. to 45-in. height typical of bandsaws on factory-supplied bases. A high table is fine for cutting small stock, but it's awkward for working with large boards being run against a fence. I made my new base as low as possible. The saw's table ended up just shy of 39 in. high, and now it's much easier to handle stock.

The base is a simple box made of Baltic birch plywood and assembled using butt joints and screws. A large drawer provides room for bandsaw blades and miscellaneous small parts. In addition to the drawer, a compartment on one side serves as a place to store the infeed and outfeed supports. The bolts mounting the casters thread into capped insert nuts. The 3-in.-wide gap above the back panel of the box allows access to the motor-mounting bolts.

The motor is simply bolted in place with its pulley in line with the pulley on the saw. The belt tension is adjusted by adding or removing sections from a link belt. Using a link belt eliminates the need for a sliding motor mount. On my saw I was able to reuse the original belt guard. If that's not possible on your saw, make a simple plywood box to cover the belt and pulleys.

The drawer slides on hardwood tracks attached to the bottom of the base. A single hardwood block attached to the top back edge of the drawer prevents the drawer from tipping when extended. The block is slightly oversize and mounted with two recessed screws. Then, for a smooth sliding fit, trim it to size with a block plane.

JOHN WHITE is a woodworker and machinist who maintains the *Fine Woodworking* workshop.

Jobs a Shaper Does Best

BY LON SCHLEINING

PATTERN CUTTING IS A SHAPER'S STRONG SUIT. To pattern-shape small parts, the author uses jigs with handles and clamps. Pattern cutting is one of four jobs where the author says a shaper outperforms a router and router table.

I'm convinced that a shaper—more than a router table—should find a home in every active woodshop. Sure, the shaper is well-suited for heavy work, like forming deep contours and complex profiles. In fact, I use the machine daily to make custom hand rails, balusters, and other stair parts. But even straight moldings and ordinary light shaping (tasks normally delegated to a router table) can be handled safely and easily by the shaper, and with better results.

I use the shaper for four jobs: running straight molding, raising panels, pattern cutting (see the photo below) and doing radius work. Each job requires different tooling and setup. When the machine is molding, for example, you'd hardly recognize it as the

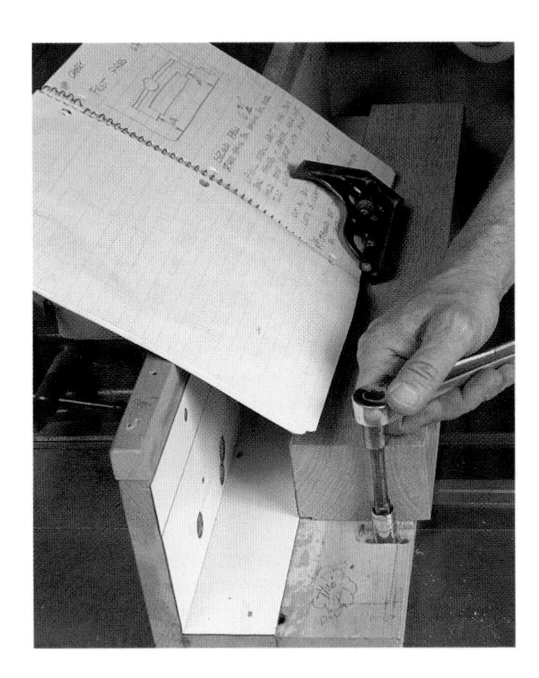

TUNNEL-SHAPED JIG IS BETTER **than a standard fence for straight molding work. Lined with plastic laminate to minimize friction, the jig guides the work smoothly because there is only a few thousandths of an inch clearance. The author's checklist is in the background.**

router turned out to be a small bit for a shaper. I still use a router from time to time, but the shaper is my tool of choice whenever possible.

Driven by a belt and dampened by lots of cast iron, a shaper just coasts through most lumber. Like a router table, a shaper has a cutter sticking up through a hole in a worktable. And many of a shaper's setups will be familiar to you if you've used a router table. But that's where the similarities end. A shaper is more solid and more powerful than a router table. Though a shaper turns at less than half a router's speed, the shaper produces a superb cut because there is less vibration. There are two reasons for this. First, most shapers weigh almost as much as a cabinet-model tablesaw (about 450 lbs.). Second, in most shapers, the cutter is fixed to a 1-in.- or 1¼-in.-dia. spindle, which is much more rigid than a ½-in.-dia. router bit shank.

TWO WAYS TO HOLD SHAPER CUTTERS. **The spindle with collet (left) holds standard ½-in.-shank router bits; the 1-in.-dia. spindle with nut holds stacked wing cutters. The carbide-tipped router bits and two-piece cutter were custom made.**

POWER FEEDER IMPROVES STRAIGHT MOLDING. **The author relies on a power feeder to run straight molding and hand rails. The feeder makes safe, even cuts. A Plexiglas guard over the cutter lets him see that the chips are being cleared out.**

same machine that raises panels. Spending time setting up each cut makes the shaper dependable and a pleasure to use.

I've gained confidence with the shaper because I do what it takes to make the machine safe (see the sidebar on p. 104). I haven't skimped on tooling, accessories or jigs. And having an assortment of cutters, guards, jigs, and a power feeder lets me shape items that I would otherwise have to buy from a millwork shop. Jigs, in particular, are great for holding and guiding small or awkward pieces (see the sidebar on p. 107).

Shaper Anatomy: More Solid than a Router Table

I've tried to do stairbuilding work using a heavy-duty router, but in the middle of a deep profile, I discovered that the router was straining to make the cut. It made me nervous routing with a 2½-in.-dia. bit that weighed several ounces. So I bought a shaper. When I put the same bit in the new machine, the cut was effortless, vibration-free, and just felt a lot safer. A big bit for a

With a Shaper, Safety Comes First

Not far from my shop there is a cabinetmaker who wears an oak apron when he's shaping. Even experienced woodworkers are edgy around shapers. But being cautious is wise. A hard thing to learn is taking enough time to be safe. When shaping, I put safety before speed and before cost.

A KILL SWITCH OFFERS SECURITY, so the author built this foot-controlled off switch. He also keeps the spindle-reversing switch taped, so he doesn't change the cutter rotation inadvertently.

DOUBLE-CHECK THE TOOLING

The biggest fear with a shaper is thrown cutters. I spoke with a guy who had to duck behind his tablesaw when the piece he was shaping kicked a knife loose. As it enlarged the hole in the shaper's top, he said it sounded like a 747 coming in on its belly. Fearing a fire from all the sparks, he slithered back over to the machine to turn it off. From that story, I've learned to do three things to minimize the risk of loose tooling. First, I don't use slip knives. I use only wing cutters, safety cutters, or router bits. Second, I recheck the tightness of every cutter I install. Third, and most important, I take light cuts while feeding the stock slowly.

MODIFICATIONS ADD A SAFETY NET

I added some extra safety features to the machine when it came out of the crate. I added a cord with a plug and did not wire the machine directly to a circuit. I keep my shaper unplugged, except when I'm running it. When I'm changing cutters or have my hands in the shaper's innards, I drape the disconnected cord where I can see it.

I made a foot-operated kill switch, which is a hinged paddle that contacts the off button (see the photo above). I can hit the paddle while keeping both hands in position, my body upright and my eyes on the cutter. Another improvement was tensioning the shaper's belt, so it will slip if a workpiece gets jammed.

USE THE RIGHT SETUPS AND STAY FOCUSED

Making the job comfortable is one shaper-safety item that's frequently overlooked. Besides wearing eye and ear protection, I make sure I have good footing. I collect old rubber door mats to use as non-skid pads.

I always pick the appropriate spindle speed for the cutter diameter (large cutters require slower rpms). Where possible, I shroud the cutter with a guard or a power feeder. If I'm using the fence, I keep the gap in it as small as practical, and I use a table insert ring sized to the cutter.

When shaping, I keep my hands well out of the cutter's reach. Because I always use either a jig or a starting pin, I am never free-handing work into the cutter. When feeding stock, I shape end grain first. I work against the cutter rotation (unless I'm climb-cutting with a feeder), and I stay out of the line of a kickback. I do not shape stock that has knots or pieces that are too short or too thin.

During shaper setup and use, I keep the shop door locked and the phone answering machine on. When my attention is drawn away from my work, I write down the next step and tape the note to the machine before I take care of the problem. When I return, I take a few extra moments to re-focus, and I don't hesitate to postpone a tricky or unfamiliar job if something doesn't feel right. That's usually when I'm about to make a mistake. –L.S.

Unlike a router table, where a router is inverted, the shaper is designed to be used with the cutter sticking up. In a router table, the motor sits directly below the cutter where lots of dust goes.

In a shaper, the motor is off to the side. Both my router and shaper are rated at 3 hp, but the router motor will develop the rated horsepower only in a theoretical scenario where power is measured in terms of wattage; the shaper delivers 3 hp at continuous speed and torque.

A shaper's spindle bearings, which are separate from the motor, are much larger than a router's, so the shaper will feel much more solid and stable. Another plus with the shaper is that the cutter rotation can be reversed, so cutters may be inverted in certain situations (I'll tell you more about that later). Also, with a shaper, you can move the spindle up and down with a handwheel and then lock the setting. This makes tiny height adjustments easy and precise—something that's difficult to do with a router table.

Accessories and Tooling Increase a Shaper's Capabilities

My Powermatic shaper has a single-phase 220v motor and two speeds: 7,000 and 10,000 rpm. The machine's 30-in.-wide table is thick enough that I can drill and tap holes in it to mount jigs and a power feeder. The shaper came with an adjustable fence with a dust port. The fence is split, so the outfeed and infeed sides can be offset, like a jointer's tables. This is essential when you're removing the entire edge of your material.

I rarely use the fence alone because I like to bolt on an auxiliary fence for most operations. The machine also came with a miter gauge that runs in a slot, like a tablesaw's, but I never use the miter gauge because I prefer using a fence.

My shaper has three spindles: a solid 1-in. spindle and stub spindles, ½ in. and ¾ in. dia. I use the 1-in. spindle for heavier work, the ¾-in. spindle for smaller cutters, and rarely, if ever, the ½-in. spindle. Wing cutters (with either three or four wings) or safety cutters (also called antikickback cutters) will slide over the spindle. I also can stack a combination of these cutters, spacers (collars) and shims to produce complex profiles. A keyed washer and a locknut hold the cutters on the spindle.

Changing cutters is more involved on a shaper than a router, but you can buy a collet for the shaper (see the top right photo on p. 107), which lets you run ½-in.-shank router bits that interchange readily. Despite the shaper's slower speed, I've found that router bits run fine. You also can use cutterhead tooling, or insert tooling, in a shaper, where the knives are locked in the head by a setscrew, an alignment pin or a V-groove. With cutterhead tooling, you can replace and swap knives, and you can grind a blank to make a custom profile. For my work, however, I'm only comfortable shaping with wing cutters, router bits, and safety cutters.

In my shop, the shaper sits alongside a central work station, so I have ample infeed and outfeed area. I built a platform for the shaper, so it is at the same height as the work station table. Because I don't use the miter gauge, I rotated the machine 90° clockwise from its conventional position. This orientation offers better access to the controls and makes changing tooling easier. I also bolted the machine to the floor and to the work station to reduce vibration. With the shaper secured, I can apply pressure without worrying that it will move. To keep the work area and my lungs clear, I have a 1,000-cu.-ft.-per-minute (cfm) dust collector that keeps up with most of the waste. For every jig, I make a dust pickup boot from a coffee can or standard metal heating duct.

Checklists and Other Precautions

Pilots use checklists every time they land an aircraft or take off in one. I also use checklists when shaping (see the photo at left on p. 103). Remembering to tighten the spindle nut, just like remembering to drop the landing gear, is too important to leave to memory alone. In a quiet moment, I write down the sequence of an operation. I include everything from locking the height adjustment to counting the pieces after a run. Each time I make a setting that I plan to use again, I make sketches and jot down the dimensions in my notebook. When I quit for the day, I mark where I have left off.

Make Light Cuts, and Take Your Time A shaper is capable of cutting in a single pass, but I only do so when I'm using a power feeder and forming relatively modest profiles. For most shaping, I use a series of light cuts, which are easier on the machine, and they get me used to the process. Instead of taking a chance of ruining a piece by hogging all the way in one pass, I take an initial pass and then clean up with light subsequent passes.

My shaper fence adjusts outward for progressively deeper straight-run cuts (see the photo at left on p. 103). For raised panels, I elevate the cutter into the piece in stages. When pattern cutting and doing radius work, I also increase the depth of cut in steps. First I bandsaw close to my lines to minimize how much the shaper has to cut. Then I use a flush cutter followed by the profile cutter. Graduated bearing sizes let me make deeper and deeper cuts.

Straight Runs: Shape with a Fence and a Power Feeder

For straight shaping runs, I always use a power feeder. To me, a shaper isn't complete without one. The immediate benefit of the feeder is that the stock moves past the cutter at a constant rate. Chatter and burn marks are gone because the stock feeds without hesitation due to changes in hand positions, which are harmful, repetitive motions anyway.

A power feeder offers other advantages. When the feeder is set slightly askew, the stock will hug the fence. Because the wheels apply constant down pressure, there is little chance of a kickback, and boards that are bowed stay flat on the table. The power feeder hovers over the cutting area, so it shields me from flying chips (see the photo at right on p. 103). Most important, though, is that a power feeder keeps my hands far away from the cutting action.

The jig I use to form straight molding (see the center photo on p. 103) resembles a tunnel. Its opening is two or three thicknesses of paper wider than the stock I'm running. This allows 0.010 to 0.015 in. clearance, so the stock slides without binding. I line the tunnel with plastic laminate, and I lubricate it with TopCote®. The key here is to have all the blanks milled consistently. I use a portable planer to thickness the stock, and I mill a couple of pine blanks at the same time so that I can test the shaper's setup. The roof of the jig is the power feeder.

Panel Raising

The conventional way to raise panels on a shaper is to run the panel face up (see the drawings on p. 108). Panel-raising cutters are designed to run above the work. There are several reasons for this. First, a panel tongue will always fit its intended groove in the frame. Even if the panel is cupped, the thickness of the tongue will be cut just

Shaper Jigs Put You in Control

INVERTED PANEL RAISING IS SAFER. Because Schleining likes to keep work between his hands and the cutter, he prefers to raise panels with the bevel facing down. The auxiliary table and guard also shield the cutter.

USE JIGS FOR SAFE AND CONSISTENT SHAPER WORK. Schleining built a jig to shape a profile on the side of a handrail piece. Secured vertically in a holder, the piece is rotated with a handle past the cutter. The shaper is turned 90° from its normal position.

Shapers require more hold-downs, guides, and stops than other machines. I've spent half a day setting up an operation that takes just a few minutes.

To build jigs, I use Finnish birch plywood because it wears well and is strong. I use ¾-in. plywood to make jig bases. To hold a workpiece, I prefer toggle clamps because they grasp and release easily, and the tension can be adjusted. I integrate a cutter guard and a dust hookup into most jigs.

For small pieces, I make the jig oversized and put handles on it (see the photo at left above). I also make the part longer than it needs to be. To keep the work from being yanked out, I screw the end of the piece where it won't be near the cutter.

I never get tired of seeing perfect contours emerge from jigs. When I'm done with a jig, I hang it on the wall, where it's always handy.

undersize, which means that it will still fit. Second, with the panel facing up, it won't get scratched on the table. Third, the operator can watch the cutter do its work.

Despite all these good reasons, however, I prefer to raise panels face down when I can. With the cutter below the work, I feel safer. Here's how I do it:

I invert a panel-raising wing cutter and set it up so that most of it is below the surface of the shaper table for the initial pass. If the hole in the table is too small to allow the cutter to descend below the surface, I put down a plywood auxiliary table.

With the cutter inverted, the shaper must run in reverse, which means I take a few extra precautions. I position a guard well over the cutter. And I draw arrows on the jig to show feed direction and rotation (see the photo at right above). There also is the possibility that the spindle nut could loosen due to the rotation, but I prevent this by using a keyed washer under the nut. I check the nut occasionally just to make sure. Finally, I loosen the red tape I keep

Shaping with a Starter Pin

A shaper makes short work of cutting the profile on a raised panel. To start the cut, the workpiece is held against the pin (see Step 1) and rotated gradually into the cutter. The panel stays in contact with the pin (Steps 2 and 3) as it moves into the bearing. When the bearing supports the work and is no longer spinning, the workpiece may be pivoted, so it's free of the pin (Step 4).

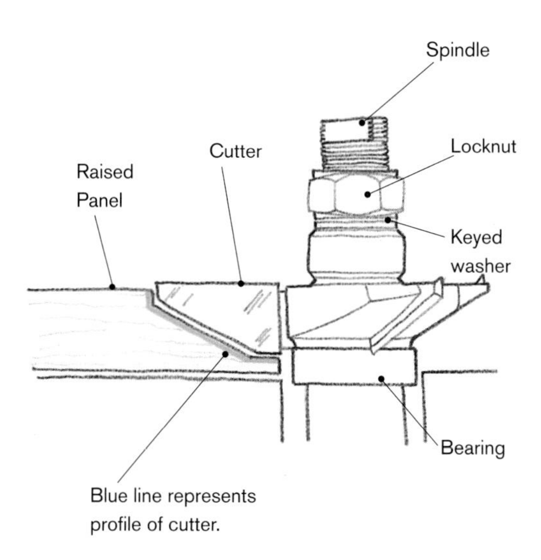

Step 1:
Panel contacts pin away from cutter.

Note: For clarity, these drawings show a panel being raised face up without a guard.

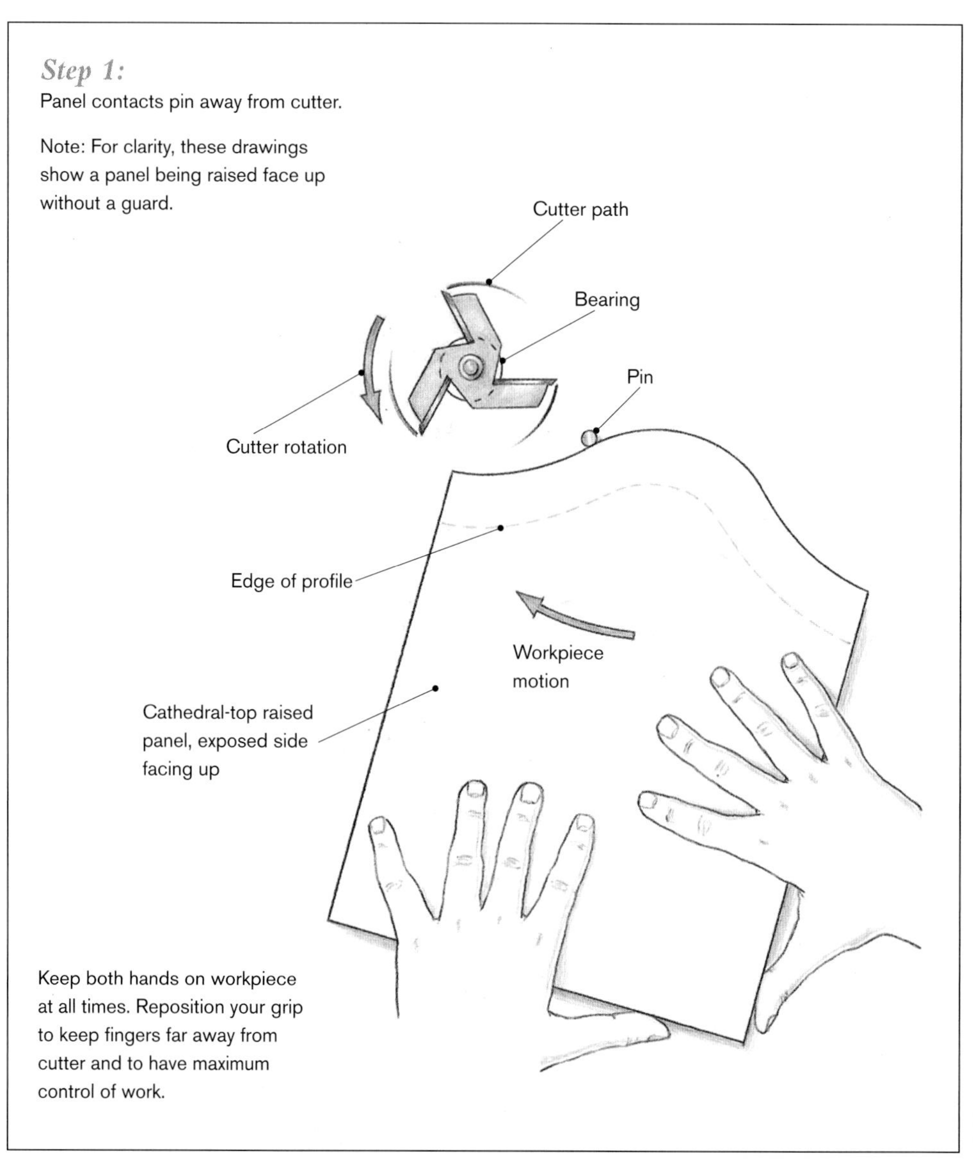

Keep both hands on workpiece at all times. Reposition your grip to keep fingers far away from cutter and to have maximum control of work.

Step 2:
Panel is held against pin. Gradually begin shaping by pivoting workpiece into cutter.

Step 3:
Panel moves into cutter until it hits bearing. Then panel pivots free of pin.

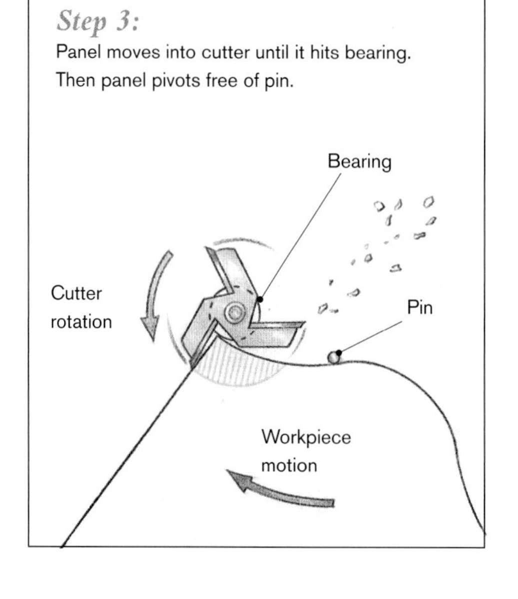

Step 4:
Panel contacts bearing and cutter only.

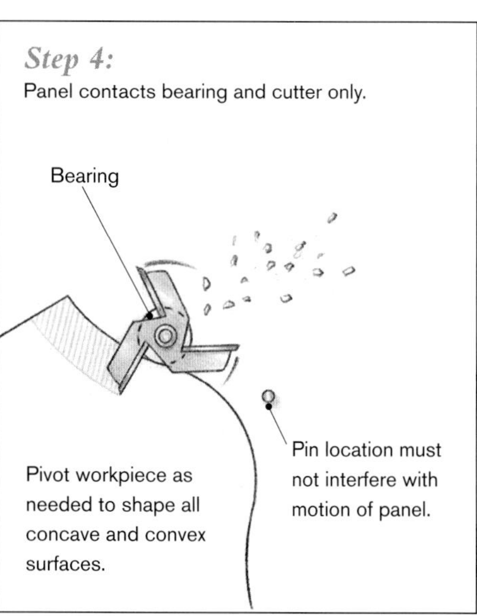

over the reversing switch and drape it to remind me that the cutter is turning opposite its normal way.

Pattern Cutting

For contouring curved parts, it's hard to beat a pattern-cutting jig and a shaper. Pattern cutting requires a guide bearing or rub collar above or below the cutter to ride along a pattern. The pattern can be the stock itself or a plywood or scrapwood template (see the photo on p. 102). As for the cutters, I usually start with a flush-cutting bit. Next I use a profile cutter and shape in stages of depth.

Using On-Ramps, Off-Ramps, and Starting Pins Usually, when I use a guide bearing, I make the pattern-cutting jig with an on-ramp and an off-ramp. The ramps are just extra pieces of wood that contact the bearing before the work and stay in contact after the work has exited the cut. The ramps allow the cutter to ease into the work and exit it smoothly without abrupt transitions. Often, I'll build the on-ramp into the end of the actual workpiece. Both ramps need to project far enough to contact the bearing while leaving the work clear of the cutter. If I can't use a jig, I use a starting (fulcrum) pin to control the cut. This pin is inserted into a hole in the table close to the cutter. The drawing on the facing page shows how a curve-top panel for a cabinet door is run using a starting pin.

Radius Work

Shaping a radius is easier than it looks. For large vertical-axis radii, I use a jig that looks like a segment of a wagon wheel laid flat on the shaper. I rotate the jig about a pivot point, so the workpiece moves past the cutter.

The setup for a horizontal-axis radius is shown in the top left photo on p. 107. The jig sits upright on the table, and the pivot point is actually above the cutter. I have slotted holes in the workpiece holder, so I can slide it closer to the cutter for gradually deeper cuts. An outboard fence on the right side of the jig prevents the work from being pulled out of its holder.

The Payoff

One day, when I was using a horizontal radius jig, the cutter somehow dislodged the piece of oak I was shaping. The motor was running, even though the cutter had stopped. I held onto the jig with both hands. Reaching out with my left foot, I hit the paddle switch, turning the machine off. I breathed again. The jig and my safety precautions had paid off.

LON SCHLEINING teaches woodworking seminars around the country. He is the author of *The Workbench* (Taunton, 2004).

Choosing Shaper Cutters

BY LON SCHLEINING

WING CUTTERS CAN BE STACKED and the tips staggered so only half are cutting at a time. These custom cutters are made with a shear angle to give a smoother cut.

A shaper is an indispensable tool in my custom stairbuilding shop. But I had used one for 12 years before I could bring myself to try anything other than standard wing cutters. I'd heard too many stories about knives that had been thrown from old-style cutterheads to feel comfortable with them on my own shaper. Eventually, I was convinced to try modern cutters with replaceable knives—what the industry calls insert tooling. To my surprise, I found a lot had changed in cutter design.

Even with new products on the market, there are still only two basic kinds of cutters: those with permanently attached cutting edges, like the wing cutters I was used to, and those with interchangeable knives. Developments in design and manufacture have produced safer, better-performing tools of both varieties. European safety standards have led to a new family of wing cutters, and insert tooling is now offered in several styles that are a big improvement over the old-fashioned cutters, which I refuse to use.

Many manufacturers make shaper cutters, and each offers a variety of profiles. Panel-raising, flush-cutting, detailing, and molding (contouring) cutters are just a few of the common styles. There are also matched sets, like cope-and-stick cutters, for making rails and stiles in doors and paneling. Generally, the bigger and more powerful the shaper, the bigger the cutter you can use and the larger the cut you can take in one pass.

Most cutters have either a ¾-in. or 1¼-in. bore, which are the two most widely used spindle diameters, and two to four cutting edges made of high-speed steel or tungsten carbide. High-speed-steel cutters are less expensive and can be ground to a sharper edge than carbide. But I use carbide because it holds an edge much longer. This is important because every sharpening reduces a cutter's diameter. For matched cutters, all the knives in the set must be carefully sharpened to maintain the mating profiles.

Not surprisingly, the best shaper cutters are the most expensive. But I consider cost last. The expense is only for the short term. Over the long haul, I've found that high-quality cutters are a better investment.

Even with new products on the market, there are still only two basic kinds of cutters: those with permanently attached cutting edges and those with interchangeable knives.

Wing Cutters Are Easy to Set Up

Three-wing carbide-tipped cutters (see the photo on p. 114) are just about my standard tooling choice. Solid cutters run true and require no special expertise to set up because there are no loose parts. They are available in hundreds of stock contours, and you can stack different cutters on the spindle to produce complex profiles. In addition, cutters can be made in almost any custom profile (see the sidebar on p. 114).

Wing cutters usually have two or three wings, but four-wing cutters are not uncommon. Two wings are best for clearing chips quickly and for removing a lot of stock in one pass, but manufacturers prefer cutters with three wings because they are easier to balance. More wings mean more cuts per minute and, therefore, a smoother cut. The cost of a cutter is directly related to the number of cutting tips.

Wing cutters often incorporate a shear angle to improve the cut quality. This is where the cutting face is angled rather than parallel with the spindle. The shearing action slices rather than chops the wood.

Safety Cutters Help Reduce Kickback

The most recent development in wing cutters is a new antikickback design, also called safety or chip-limiting cutters. Developed in Europe, safety cutters limit how far cutting edges protrude from the body of the cutter. Viewed from above, these cutters look more like a disc than a propeller. Because the body of a safety cutter is only slightly smaller than its cutting diameter and because there is no open passage behind the cutting face, these cutters require a slower feed rate. In use, safety cutters and conventional cutters are impossible to tell apart. Kickback is minor. The biggest drawbacks to the new safety cutters are high cost and limited availability. Also, local saw shops may not be able to sharpen them.

Insert Tooling Is Versatile

Replacing just the knives rather than the entire cutterhead makes insert tooling attractive to both production shops and one-off operations (see the bottom left photo below). In a production run, knives that get dull or damaged are easily replaced, and having a spare set reduces down time. In addition, replacing dull knives with new ones maintains a consistent profile.

Interchangeable Knives

LOCK-EDGED CUTTERHEADS have interchangeable knives. Teeth in the edge of the knife engage a screw in the cutterhead.

CHIP-LIMITING INSERT TOOLING provides increased safety. A plate in front of the knife exposes only a small portion of the blade to prevent overfeeding.

WING CUTTERS USE INSERT TOOLING. Replaceable tips mean little down time when a cutter becomes dull or damaged.

WEDGE-SHAPED GIBS hold the knives in place. A pair of pins holds each knife to the cutterhead if the gib ever loosens.

For short runs in a variety of profiles, insert tooling is economical, especially when you are using high-speed-steel knives. These steel knife blanks can be ground to virtually any profile.

As a group, though, interchangeable cutters require more care in setup and use. Every fastener must be carefully tightened. Knives must be in good condition so that they will seat properly. And they must be sharpened as a set to remain balanced. Finally, the assembly must be checked to be sure that the knives protrude equally from the cutterhead.

A slip-knife cutterhead is an early example of insert tooling. The slip knives are clamped on edge between two collars, and friction caused by compression of the spindle nut is all that holds the knives.

The disadvantage, and major safety concern, with slip knives is that they can fly out of the cutterhead if the spindle nut is

Fixed Knives

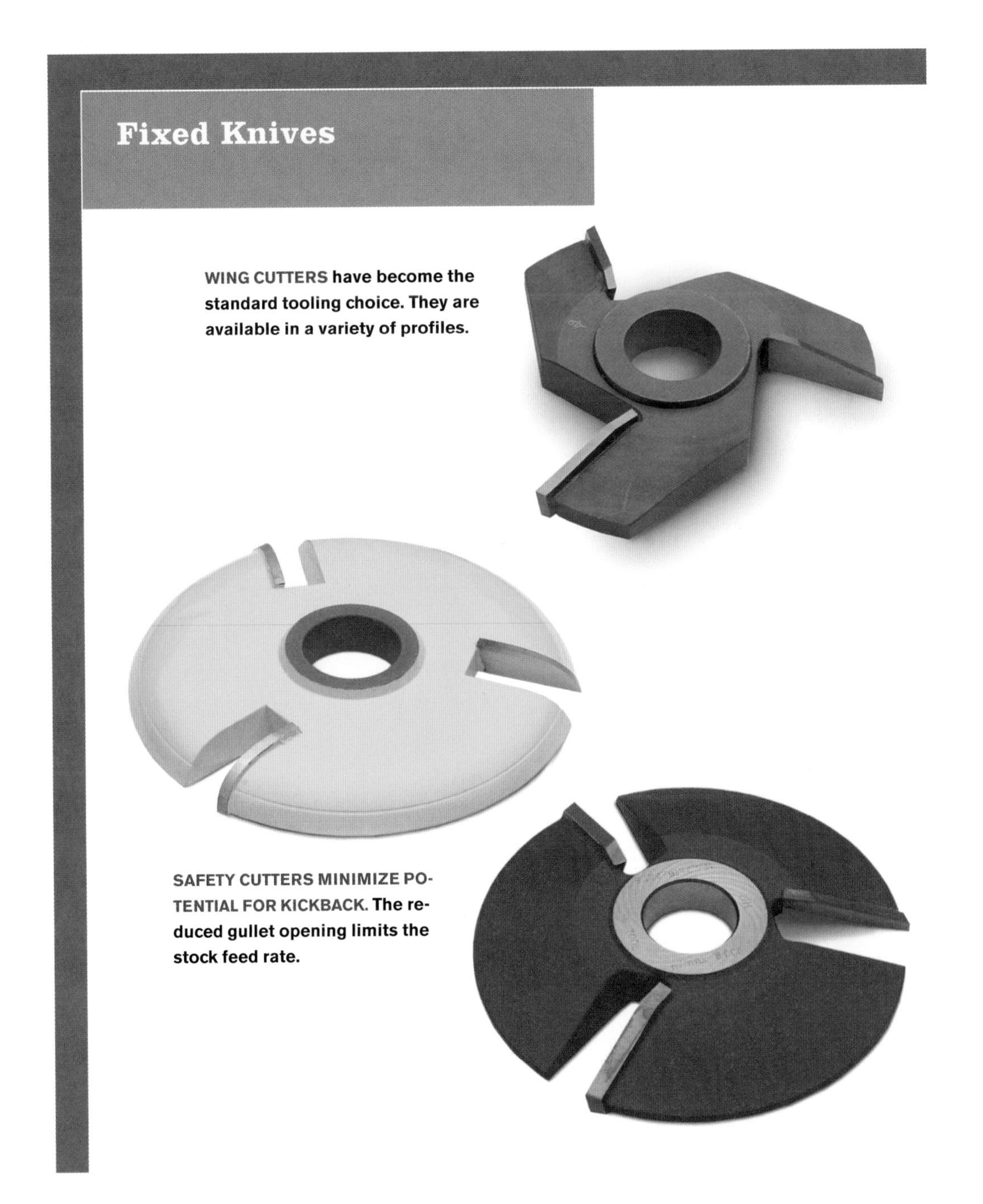

WING CUTTERS have become the standard tooling choice. They are available in a variety of profiles.

SAFETY CUTTERS MINIMIZE POTENTIAL FOR KICKBACK. The reduced gullet opening limits the stock feed rate.

Ordering Custom Cutters

CUSTOM-MADE, carbide-tipped wing cutters save time. Two cutters mold this large banister profile with the fewest setups.

Despite the availability of many stock profiles, you may need a custom shape. Some toolmakers can produce a custom profile from a drawing and have it ready the next morning.

To locate a good toolmaker, call the supplier where you get stock cutters or ask at the blade-grinding shop that you use. A knowledgeable salesperson can save you time and money.

It helps to remember that a toolmaker is first a metal worker. He may know a lot about tooling for woodworking, but woodworking may not be his area of expertise.

The more information you can supply, the better. Specify the basics: cutterhead type, knife-blank size, spindle diameter, and machine horsepower. It also helps to know the material being worked and whether it will be hand- or power-fed.

A careful drawing is a good start. If you are trying to match a pattern exactly, send a sample of the molding itself. If you are cutting a curve, specify the radius of the curve and how you intend to run the stock through the shaper. When in doubt, ask for a return drawing with your quote so that you can see if the cutter the toolmaker has in mind is actually the tool you need. And if he has made something similar for another client, a call to that person may be a good idea.

–L.S.

not sufficiently torqued down. You have to be especially careful when you install and use these cutters. Setup is time-consuming and requires a higher level of expertise than any other type of tooling.

Latest Insert Tooling Adds Safety to Versatility

A safer alternative to older style slip knives are lock-edged knives (see the top right photo on p. 112), which have milled teeth in one edge of each knife. These teeth engage an adjusting screw in the collar. This forms a mechanical connection between the cutterhead and the knives and allows the knives to be set so they all have a consistent cutting diameter. Some toolmaking shops can mill teeth in the edges of older slip knives so that they can be used in a lock-edged cutterhead.

Another type of insert cutterhead holds a pair of high-speed-steel knives in place with wedge-shaped gibs, which is somewhat similar to the setup for some jointer knives. A pair of pins registers each knife in the cutterhead and forms a mechanical safety lock (see the bottom right photo on p. 112).

A similar design also uses a wedge-and-gib screw but adds serrations across the back of the knives. These grooves interlock with matching grooves in the cutterhead to form a mechanical connection. The serrations also provide a reference to help set the distance the knife projects from the cutterhead.

Insert tooling can be purchased with a full set of knife profiles. Or you can purchase just the cutterhead and the knives you need. A number of manufacturers offer pre-ground knives and blanks made of high-speed steel. In addition, some manufacturers sell solid-carbide knives.

LON SCHLEINING teaches woodworking seminars around the country. He is the author of *The Workbench* (Taunton, 2004).

Boring Big Holes

BY ROBERT M. VAUGHAN

Many years back, I had a commission to build lockset displays. The job called for 18,000 holes with smooth bores and crisp edges in blocks of 1¾-in.-thick oak. At a flat rate of 9 cents a hole, I couldn't afford to let the wrong drill bits slow me down.

Before beginning, I experimented with a variety of methods. I settled on Multispur® bits chucked in a drill press. They produced precise, tearout-free holes and allowed me to work fast enough that I finished the job in a little over a week. Depending on the project, other large-hole boring tools might be worth considering. The most common tools include Forstner bits, spur bits, spade bits, hole saws, and wing cutters.

Furniture, craft projects, architectural work, and home repairs often call for boring large holes. Large in my book is anything more than ⅝ in. dia., bigger than most commonly available twist drill bits. Big holes demand special bits, and the variety on the market includes everything from inexpensive high-carbon-steel spade bits to costly Forstners.

Before spending a wad of cash on a set of bits, consider how often you need to drill large holes, the precision of cut

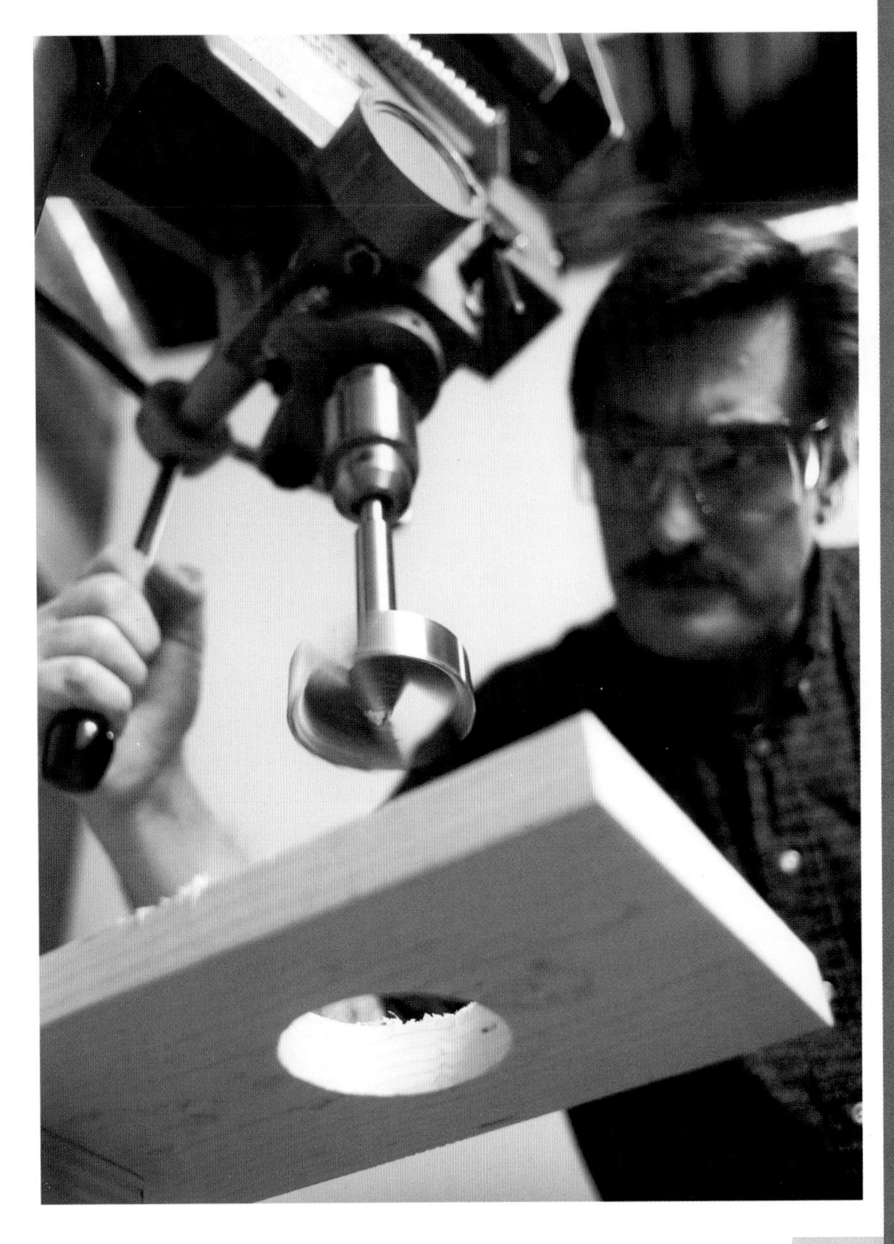

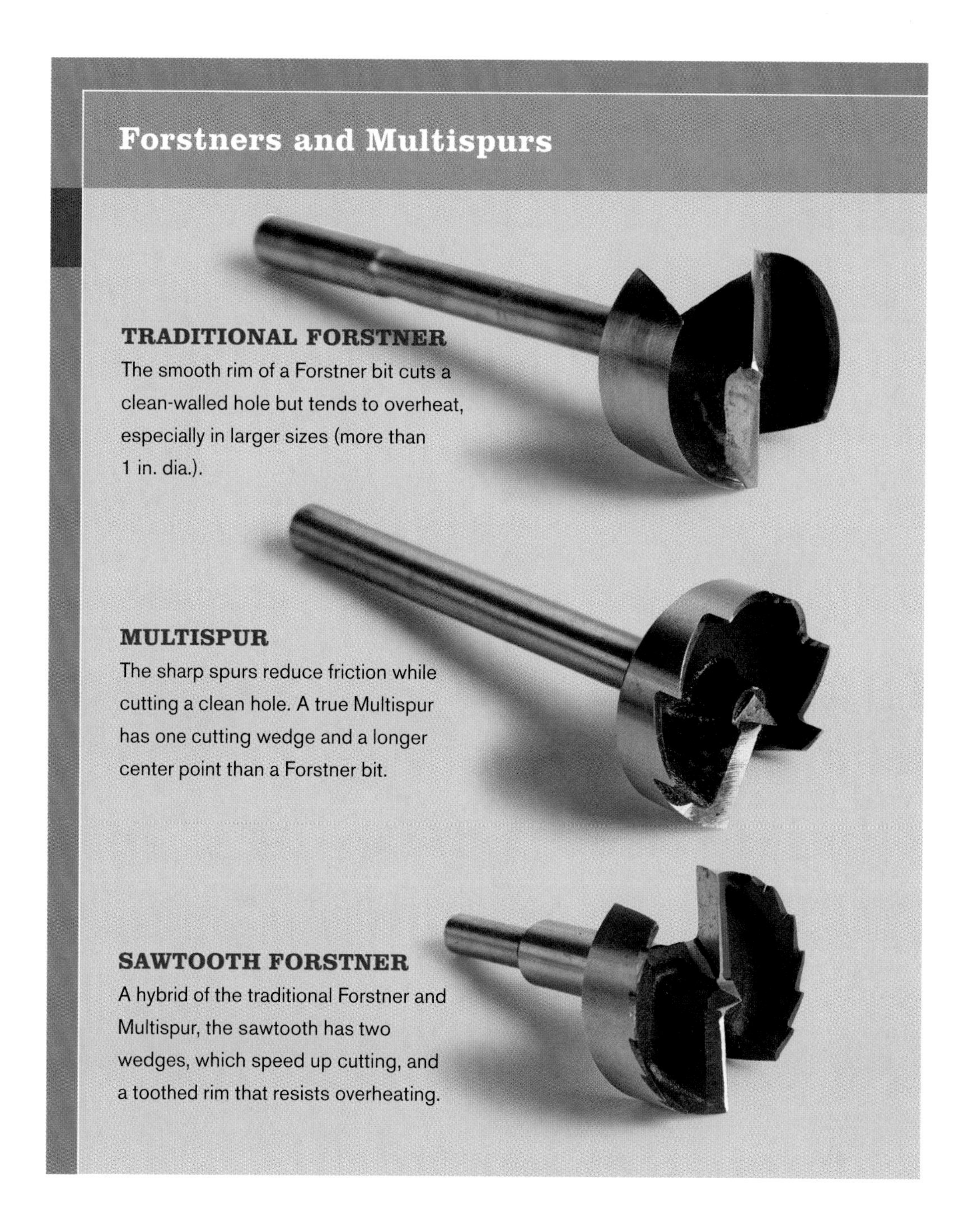

required, and how quickly you need to get the job done.

For Precision Holes, Pick a Forstner or Multispur Bit

Like life forms, tools evolve over time, only much faster. Forstners were developed more than 100 years ago for use in hand braces. They were an improvement over the other bits of the time, such as brace bits, because Forstners could drill overlapping and flat-bottomed holes. Forstners cut on two fronts: A sharp outer rim continuously scores the wood, and a pair of horizontal cutting wedges removes most of the waste inside the hole and shaves the bottom flat.

If brace bits were life forms, they'd be the fish of boring tools. Forstners are the amphibians. The 20th century saw yet another major evolution: Forstners emerged from the swamp with teeth along their rims. Although it would seem that this would give the bit a bigger bite, something more significant happened. Like mammals, these sawtooth Forstners were more efficient at heat regulation. Getting rid of the solid rim meant less metal-to-wood contact, which creates heat-producing friction. With the advent of power tools, a cooler-running bit was needed. Smooth-rimmed Forstners, especially those 1 in. and greater in diameter, tend to scorch wood when used in an electric drill.

Today, we can choose among traditional Forstners, sawtooth Forstners, Multispurs, and spur bits. All do the same thing: They drill flat-bottomed holes with clean rims and smooth sides. (Although a traditional Multispur has only one cutting wedge, the term multispur has become synonymous with the sawtooth Forstner, which has a pair of cutting wedges.) These bits are good for boring holes for dowels or hardware, architectural detailing, or craft projects, such as clock making, where you need a precise hole.

True Forstners have a stubby center point that just barely protrudes beyond the edge of the rim. When centering the point on a mark, you have to peek under the bit to see if it's in the right place. Sawtooth Forstners and Multispurs have slightly longer center points, which make them easier to align. These bits range from ¼ in. to as big as 4 in. dia., and prices vary widely (see the sidebar on p. 118). Spur bits have a pair of cutting wedges and two small spurs.

BORING OVERLAPPING HOLES. Forstner bits are good for roughing out mortises.

SPECIALTY SPUR BITS. Spur bits have a pair of small spurs on their rim, and they're made for boring at 90° to the surface. This one is made for installing 35mm cup hinges.

They're made for boring at 90° to the surface. Many 35mm bits for installing cup hinges are of this style (see the photo at left).

Because the outer rim of a Forstner is smooth, it won't catch when drilling overlapping holes (see the photo above) or when boring at an angle. Sawtooth Forstners and Multispurs can also be used to drill overlapping holes, but be aware that the teeth on these bits can catch and hurl a workpiece that isn't clamped down. When drilling overlapping holes with

What's the Difference Between Cheap Bits and Pricey Bits?

The prices of Forstners, sawtooth Forstners, and Multispur bits vary widely depending on where you buy them, where they're made, and how long and thick their shanks are. I tried four 2-in. bits from three continents, priced from $15* to $58*, and drilled a bunch of holes in hardwood (see the photo below). Used in a drill press, all were capable of boring holes with acceptably clean rims and fairly smooth walls, although the cheapest bit cut the slowest.

So why the big price difference? The size of a bit's shank plays some part: more steel means greater cost. Lower-cost bits have 5⁄16-in. or 3⁄8-in. shanks about 2½ in. long. Higher-priced bits have ½-in. or 5⁄8-in. shanks about 5 in. long. A hefty shank provides additional stiffness, and the extra length lets you bore deeper holes.

I ordered a $15* sawtooth Forstner bit, made in China, from Woodcraft® (800-225-1153), a retail chain and mail-order company. This economy bit was slow-cutting. The rim of the hole had some minor tearout, but overall, it was acceptable. The spurs and cutting wedge weren't ground as sharp as the other bits. For occasional use, this bit would be a good value.

A $35* Forstner bit from Austria, ordered through Woodcraft, came with a sharply ground outer rim. This Forstner cuts faster than the Chinese bit, leaving a clean hole. A $46 Multispur from Forest City Tool, made in Hickory, North Carolina, cut faster than the other bits. The Multispur produced some minor tearout at the edge of the hole and on the wall. The most expensive bit, a $58* sawtooth Forstner from Austria ordered from Woodcraft, cut as fast as the U.S.-made bit but left a tearout-free rim and a clean wall.

–Anatole Burkin, Editor

BITS FOR EVERY BUDGET 1. Chinese-made, short-shank Forstner ($15) cuts slowly but produces an acceptable hole with only minor tearout at the rim. 2. Austrian-made, short-shank Forstner ($35) drills a cleaner hole. 3. U.S.-made, long-shank Multispur ($46) bores faster but leaves minor tearout. 4. Austrian-made, long-shank sawtooth Forstner ($58) cuts as quickly as the U.S.-made bit but leaves a cleaner hole.

these bits, use a drill press and firmly clamp the workpiece.

For Fast Drilling, Use a Spade Bit

A spade bit is one of the simplest drill bits. It's nothing more than a rod of steel with one end forged flat. The flat section is ground away, leaving a sharp point and a pair of cutting wings. Spade bits are mostly used in construction with portable drills. These bits bore holes quickly but tend to wobble and vibrate, causing a raggedy rim on the entry side and tearout and splintering on the exit side. But when you're drilling holes for wire or pipe in studs and joists, speed matters more than looks. For deep holes, withdraw the bit frequently to clear the chips, so the bit won't bog down.

When sharp and used at slow speeds in a drill press, a spade bit can cut a fairly clean hole. Spades are inexpensive and simply designed, which makes them a good choice if you need to reshape one for a special job. If you need to drill a hole to accept a 1-in. dowel that's sized a hair under 1 in., just file away an equal amount on both outside wings of a 1-in. spade bit to match the dowel's diameter (see the top right photo). Or, if you want a slightly tapered hole, you can file a taper onto the outside wings. Spade bits range from ¼ in. to 1½ in. dia. and cost about $2* apiece.

FOR SPEED, CHOOSE A SPADE BIT. **Holes for threading pipe or conduit in framing lumber don't have to be pretty.**

SPADES ARE EASY TO CUSTOMIZE. **Just file away equal amounts on each side of the blade to create a narrower or tapered bit.**

Hole Saws Are Good for Installing Locksets

A hole saw, like a spade bit, has limited use in furnituremaking, but it can be handy around the house. Hole saws come in different styles, but they generally have an arbor that holds both a pilot twist drill (usually ¼ in.) and a larger cutting cylinder, ⅝ in. or larger (see the photo at left). More expensive hole saws come in kits with an arbor that accepts hardened steel cylinders of various sizes. Costs are based on the size of cylinders; for a good arbor and a 2-in. bit, expect to pay about $20*. The cutting cylinder has fine teeth, like those found on a hacksaw. Inexpensive hole saws have thin-walled cylinders that flex like a hacksaw blade, and these sometimes pop off the arbor. They cost about $10*.

HOLE SAWS HAVE SELF-CENTERING PILOT BITS. **They make quick work of drilling out doors for locksets. You can use them in a portable drill or drill press.**

Hole saws are commonly used in portable drills by plumbers and electricians for boring pipe and conduit holes. They're good for drilling holes for locksets in doors. When using a hole saw, the pilot bit enters first, followed by the cutting cylinder. For a relatively clean entrance hole on two sides, the drill is stopped as soon as the pilot punches through the door. The hole saw is withdrawn, and the hole is completed by drilling from the other side. A hole saw doesn't leave a big pile of shavings like a Forstner; rather, most of the waste is in the form of a cylinder stuck to the pilot bit. Only through-holes can be bored with a hole saw. The surface left by a hole saw is fairly rough, so it's not a good choice when the cutout will be exposed or if you need a precise fit.

Wing Cutters and Circle Cutters Are Adjustable

Wing cutters, like hole saws, have a center pilot bit. But instead of a cutting cylinder, a wing cutter has a single vertical cutting blade attached to an adjustable bar (see the photo at right). The adjustability has some appeal. If you only need to drill a few large holes, a wing cutter, which can be purchased for about $20*, is cheaper than a large Forstner or Multispur bit. Some wing cutters can cut holes up to 6 in. dia. That makes them useful for drilling out holes for large fixtures such as light canisters. You can also use a wing cutter for cutting out wheel blanks used in toys.

A wing cutter works slower than most hole-boring bits. Its shape makes it prone to vibration, and the rotating cutter arm can give you good cause for anxiety. A wing cutter bores by cutting a groove in the workpiece, leaving a disc or cylinder. You can adjust the cutter to bevel either the waste disc or the hole. Wing cutters must be used at very slow speeds (250 rpm) in a drill press, and the work must be clamped. There are a variety of wing cutters on the market. The ones marketed as wheel cutters leave a more desirable profile on the disc.

A WING CUTTER GIVES YOU INFINITE ADJUSTABILITY. **The single cutting wing can be positioned anywhere along the bar and is held in place with a setscrew. Properly sharpened, a wing cutter can make a clean entry hole, but the walls will have some tearout. Use a drill press running at slow speeds with this tool.**

* Please note price estimates are from 1998.

ROBERT M. VAUGHAN is a contributing editor to *Fine Woodworking*, He repairs and restores woodworking machinery in Roanoke, Virginia.

A New Curve in Drill Bits

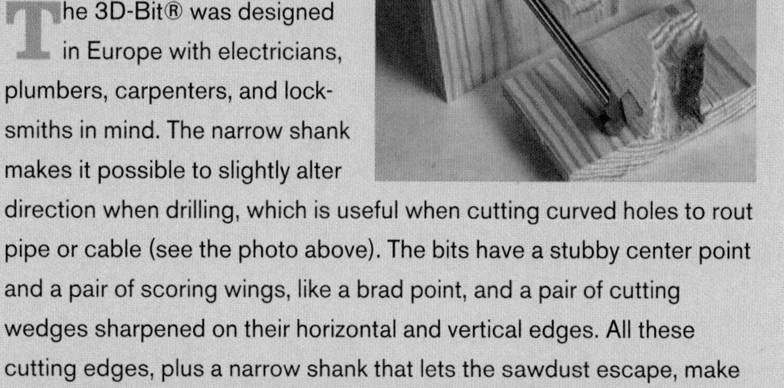

The 3D-Bit® was designed in Europe with electricians, plumbers, carpenters, and locksmiths in mind. The narrow shank makes it possible to slightly alter direction when drilling, which is useful when cutting curved holes to rout pipe or cable (see the photo above). The bits have a stubby center point and a pair of scoring wings, like a brad point, and a pair of cutting wedges sharpened on their horizontal and vertical edges. All these cutting edges, plus a narrow shank that lets the sawdust escape, make these bits fast-cutting. The cutting wings of a 3D-Bit score the wood, much like a brad point, leaving a clean entry hole. Used with a drill press, the 3D-Bit bores a hole nearly as well as a Forstner bit. 3D-Bits are being marketed by RotoZip Tool Corp. (800-521-1817). Currently, they are only available in metric sizes. They cost about $75* for a set of four.

–A.B.

Jigs for the Drill Press

BY GARY ROGOWSKI

Like most power tools, the drill press won't tackle too many woodworking jobs without jigs to hold work safely and securely. I make all of my jigs out of wood and wood products such as plywood and medium-density fiberboard (MDF). I make the jigs as simple as can be and use them to handle stock of odd shapes and sizes and to bore at any angle.

The drill press is primarily designed for metalworking. Its metal stock table is too small for clamping large boards. So the first order of business is to add a larger auxiliary table made of MDF or plywood. A simple solution is to screw the auxiliary table to the stock one. Or if you prefer a table that's fast to remove, make one that can be clamped to the metal table.

Every Drill Press Needs a Fence

When drilling a large hole, a bit can grab a board and turn it into a spinning weapon. Unless you enjoy getting slapped around by lumber, keep a fence clamped to your drill-press table. Even if stock isn't butted right up to the fence, it still provides a measure of safety because it will stop sudden rotation of a workpiece.

A fence is a must when you need to drill multiple holes a set distance from the edge of the stock. The only critical adjustment is the distance from the center of the drill bit to the edge of the fence. Clear away chips from the edge of the fence when registering stock against it. And use a straightedge to check your fence regularly to make sure it hasn't warped.

Use Stop Blocks when Drilling Multiples Whenever you must drill more than one of something, use stop blocks to register stock. The method is faster and more accurate than marking individual pieces. A stop block is nothing more than a piece of wood clamped to the drill-press fence. I also have a shopmade tilt-up stop that I can move out of the way, but not so far away that I misplace it (see the sidebar below).

For drilling multiple holes in a workpiece, such as when drilling shelf pins for a bookshelf or cabinet, I use a series of spacers to register stock. Line the spacers up along the fence, registering the first one against a stop block. Position the stock against the last spacer, drill a hole, then remove one block. Repeat. I have a stack of different-sized blocks within easy reach of my drill press.

Two Ways to Cut Mortises on the Drill Press

Before I owned a plunge router, I used my drill press for mortising. A brad-point bit will do a pretty good job of establishing a neat row of holes that can be cleaned up with a chisel (see the bottom photo on facing page).

Repetitive, Accurate Drilling

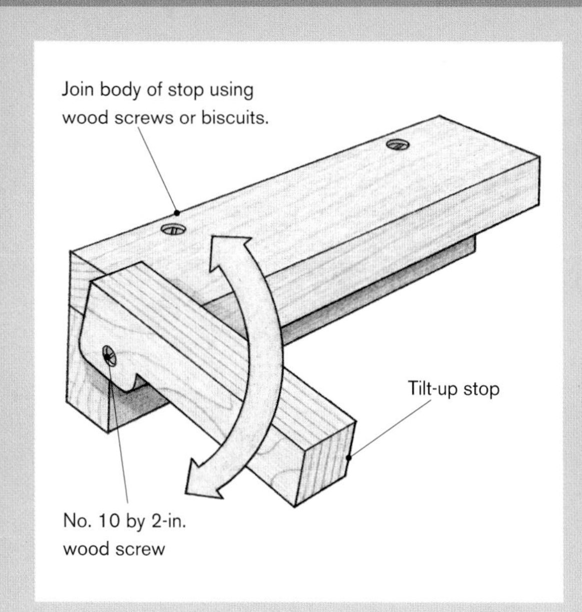

Stop blocks, either hinged (below left) or in the form of spacers (below right), guarantee accurate results when boring multiple pieces or a series of holes.

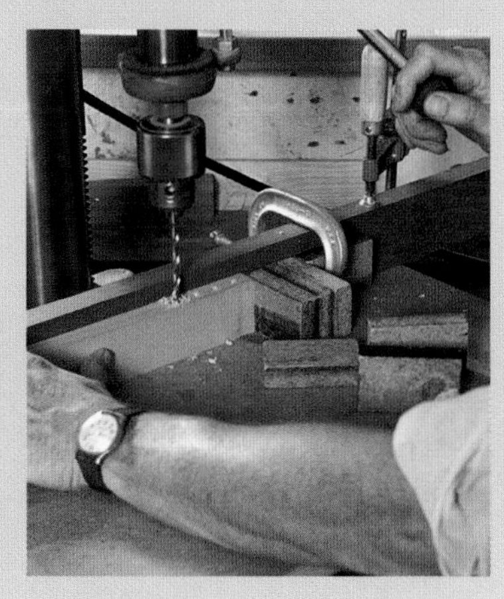

Use a straight fence and stops to locate both ends of the mortise. Drill the two outside holes first and then work your way down the mortise, overlapping holes a little. Leave some wood for the brad-point center to bite into; otherwise, the bit will drift.

I also made a sliding table for mortising on the drill press. The table has two parts: a movable sled, which is fitted with a pair of runners, and a base, which has grooves for the runners and is bolted to the drill-press table (see the photos and drawing on p. 124). The sled is made up of a double layer of glued-up material, thick enough to plow grooves for the runners, which are glued in place, without weakening it.

The sliding table has a fence and requires a stop block to locate the start of the mortise. I also clamp a stop block to the underside of the sled to control the length of the mortise. To use the jig, hold or clamp stock in place and use an end mill, a metal-working bit, to bore the mortise. Take light passes. If it chatters, switch to a brad-point bit, smaller in diameter than the end mill, predrill a series of holes, and clean up the walls of the mortise using the end mill.

Nonsquare Stock Must Be Held Firmly

Once in a while you'll need to drill stock that isn't flat or square. Bowling balls come to mind, but that's another article. Cylindrical stock can be held using a V-shaped block, which provides two-point contact and plenty of stability (see the photos on p. 125). To make a V-block, rip a groove on one side of a thick piece of wood, such as a 2x4, using the tablesaw with the blade tilted 45°.

For other shapes, you just have to improvise. Wooden screw clamps are good at holding oddly shaped pieces. Clamp the wood screw to the drill-press table, then clamp the stock to be drilled in the screw clamp. Err on the side of more rather than fewer clamps if you have doubts.

A Bigger Table

To provide a larger working surface, clamp an auxiliary table made of plywood or MDF to the stock drill-press table.

Basic Mortising

By trapping stock between two stop blocks, a mortise can be roughed out using a brad-point bit.

Tilt the Stock when Drilling at Angles Other Than 90°

Most drill-press tables tilt along one axis. But I am admittedly lazy, and I don't like moving my table back and forth and retruing it to 0° if I can avoid it. Plus, the angle gauges that come with most drill presses leave a lot to be desired.

Mortising Jig

The jig slides back and forth on runners. Using an end mill (a metalworking bit), the author takes light passes to cut a mortise.

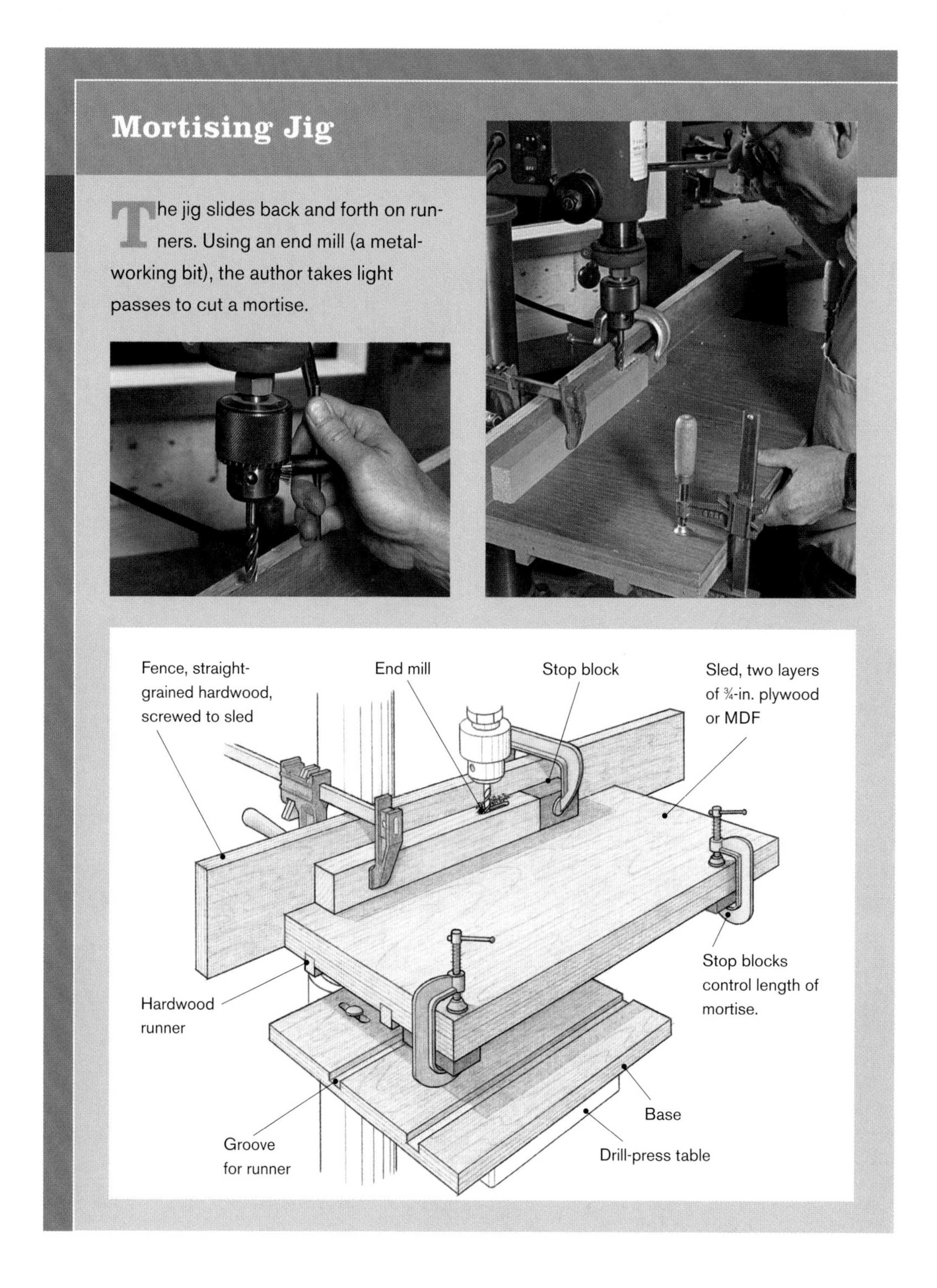

I have found that the simplest way to drill angles other than 90° is to tilt the stock, not the drill-press table. The first step is to mark the desired angle onto the stock. Then place a piece of scrap wood under one end of the workpiece. You may have to move things around until the layout mark is in line with the drill bit. Use a square or triangle, if needed. Before drilling, be sure the workpiece is stable.

A more stable angle-drilling jig can be made by joining two pieces of plywood with a piano hinge (see the photo at far right on the facing page). By wedging a wood block between the two plywood pieces, you can reach the desired angle.

Or better yet, screw the block in place so that it won't creep on you.

A Dedicated Angle Jig for Drilling Pocket Holes There are a lot of ways to attach a tabletop. One method is to run a screw through a pocket hole drilled on the insides of the table's aprons. I drill these pocket holes using a dedicated tilted fence on the drill press. I made the fence of solid stock and ripped one face at 15° on the tablesaw.

To drill the apron, hold or clamp it against the fence. Use a standard twist-drill bit when drilling at an angle, although a Forstner bit would also be appropriate. Feed the bit slowly to prevent it from grabbing.

Compound Angles There are two types of compound angles: equal and unequal. Equal is just that; both angles are the same. But chairs are rarely that simple. For example, a stool leg may hit the floor at an 80° angle from one side and 82° from the other side. That's an unequal compound angle.

Compound angles force me to tilt the drill-press table. That gets me the first angle. The second angle comes by way of a piano-hinged jig. As a precaution, place layout marks on the stock and double-check them before boring away.

Use a Two-Part Jig to Drill into End Grain

Drilling into long boards requires one of two things: great patience or another indispensable jig. You can simply tilt your drill-press table to 90° and maneuver the stock into position and clamp it. That usually entails a lot of fiddling.

Here's a better way. Make up a vertical two-part drilling jig (see the photo and drawing on p. 126). The jig is similar to the mortising jig in that it consists of a base and a movable sled with a fence. Stock clamped to the fence and the workpiece can be moved fore or aft and remain plumb (or at whatever angle the jig was set to).

Just like a tablesaw, the drill press can handle a lot of jobs in the workshop, but the machine demands a host of jigs before it truly performs to capacity.

GARY ROGOWSKI is a contributing editor to *Fine Woodworking.* He runs The Northwest Woodworking Studio, a school in Portland, Oregon, and is the author of *The Complete Illustrated Guide to Joinery* (The Taunton Press, 2001).

Jigs for Round Stock or Angled Work

The V-block can be made on a tablesaw by ripping a groove in thick scrap with the blade set at 45°.

Connect two pieces of plywood with a piano hinge. Fit a wood wedge between the leaves to create the angle needed.

Vertical Boring Jig

For boring into end grain, an adjustable table and fence provide a solid clamping surface. Wedges may be placed between the stock and base of the drill press for additional stability.

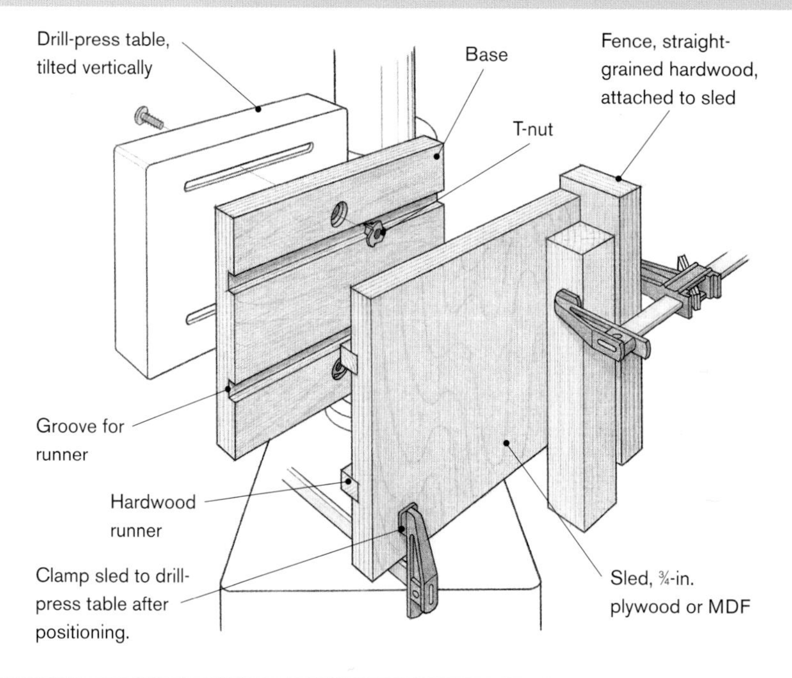

Buying Used Machinery

BY ROBERT M. VAUGHAN

I've been buying and repairing used woodworking machinery for more than 20 years. Demand for good used equipment is stronger than ever, and woodworkers often ask for advice on what to buy. I always tell them the same thing: The condition the machine is in matters more than the price. You can save money buying a used machine just as you can buying a handyman special when shopping for a house, provided you're handy and you go into it with your eyes open.

Every woodworker has his own repair tolerance. Some are intimidated by changing a belt, and others have no qualms about making new parts or welding frames. You'll do yourself a favor by being honest about your limits before buying a used machine.

The price of used equipment depends on its condition and desirability. An essential machine, like a tablesaw, in good condition often sells for close to the brand-new price. A less-desirable machine from a little-known manufacturer may sell for a third of its original price.

Just because a machine is old doesn't necessarily mean it's better than a new one. Some new models have compelling advantages, such as better safety features, easier cutter adjustments, or a greater assortment of accessories. But a used professional-grade machine will probably perform better than a new entry-level model. Will a used machine need work? Count on it.

A thorough knowledge of new machine prices and potential repair costs will help you decide when a used machine is worth the effort to repair it. A big bandsaw with a three-phase motor may need $1,000 in electrical parts alone. If you had planned to buy a $15,000 bandsaw, the investment would be worth every penny. But if a modest 14-in. bandsaw selling for $200 will need $500 in parts, it's clearly no bargain.

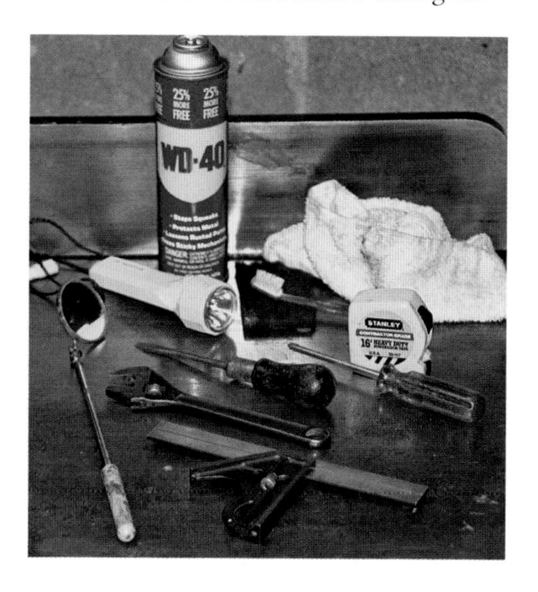

A GOOD INSPECTION requires some disassembly. Be sure to take along the tools to do it right.

Finding Good Used Machines

Finding good used equipment is often a matter of waiting. For me, the best deals seem to come from letting plenty of people know what I'm looking for. I sometimes end up with dead-end leads. But I also learn about some great buys before they get into the classifieds.

Auctions are the next best source for machinery, but the bidding can be fiercely competitive if a lot of dealers are present. Bankruptcy auctions are best—the machinery was probably being used until the time of the sale. Government or school auctions are sources of good equipment, but you may have to replace expensive, broken parts. Surplus auctions have the highest risk of bad machinery. Much of what's offered is being unloaded by large companies that have found the machines too expensive to repair.

Reputable machinery dealers are always a good source. They know tools and won't handle worthless junk. The tools are sold as is, which usually means they need some work. But the dealer may be a good source for parts and will have the means for moving or shipping the machine. Most dealers are accustomed to working with buyers from production shops who know exactly what they want. Dealers can be less tolerant with picky small-shop owners. You'll endear yourself to a dealer by doing your homework and acting decisively.

Take Along a Buyer's Tool Kit

I carry a small tool kit when checking out a machine (see the photo at left). I take a flashlight and a mirror to get a good view of obscure areas, and I take a combination square to check surfaces to see if they're square, flat, and straight. For removing guards and covers, I carry slotted and Phillips screwdrivers and an adjustable wrench. An old toothbrush, a rag, and a can of WD-40 are useful for cleaning away rust and grime. And, of course, using a tape measure is better than guessing whether the machine will fit through the shop door.

It's impossible to memorize parts and price lists for every machine that you intend to check, so take several manufacturers' catalogs with you for reference. These catalogs are available (sometimes at a cost) from hardware stores, machinery dealers, or the manufacturer. Good catalogs picture the complete machine and often give the prices of major accessories such as starters, motors, and other parts that may need replacement.

A Well-Known Brand Is Usually a Good Buy

An inexpensive, entry-level machine should be carefully scrutinized because it may have been misused. Chances are the owner was a beginner who may have used the machine improperly. However, long-defunct brands of small woodworking machines can be good buys. Many machines by Darra James, Duro, J-line, King Seeley (Sears), Red Fox, Shopmaster™, Shopmate®, Walker-Turner, and Wallace are still in service. I would buy a machine made by one of these manufacturers as long as there were no missing parts that couldn't be easily replaced (see the top left photo on p. 131). Off-the-shelf, aftermarket parts like belts, pulleys, motors, fences, miter gauges and blade guides are easy to replace. A missing jointer fence or a lathe tailstock is another matter. These are expensive parts, available only from the manufacturer. Replacing one can cost more than an entire yard-sale machine.

Companies like Oliver, Northfield®, Newman-Whitney, and Tannewitz® make large industrial machines that aren't as commonly known as Delta or Powermatic. These heavyweights handle big stock easily, but replacement parts are not cheap. Just a service manual, for example, can cost more than $50★.

KNOW YOUR LIMITS FOR REPAIRS. This 1948 Delta Unisaw cost $450. It took $150 worth of parts and 40 hours to restore this machine to like-new condition.

Check the Adjustments First

Once I've determined the machine has all its parts, I have a close look at all the adjusting mechanisms. I raise and lower the blades and tables, tilt the fences and blades, and run the guides up and down, checking for stiffness or sloppiness that will take time and effort to repair. I make sure tables and fences are square or can be made so (see the bottom left photo on p. 131). I look for damaged screw or bolt heads that may indicate the owner tried to fix the machine and decided to dump it when the repair didn't work.

With this inspection, I get a feel for how the tool was used and maintained. A machine with signs of neglect is probably no bargain (see the top center photo on p. 131). It will probably have many parts that will need to be replaced.

When I see extensive rust, I walk away. It usually means the machine has been in a fire or a flood, and those things mean big problems. The heat of a fire can distort and soften critical parts. The silt in flood waters penetrates every part of a machine, so it will have to be totally disassembled and cleaned. It's likely that every bearing will have to be replaced. Rust can also make disassembly very difficult.

Check the Frame and Table for Distortion

With a flashlight and a mirror, I carefully inspect the frame and the table for damage. Ideally, they should be in sound condition, but I don't automatically dismiss a machine with a cracked iron casting. Castings can be brazed or welded. In my area, a welded repair to an iron casting costs about $30. If I see evidence of a repair, I make sure that it didn't warp or distort the metal and that the parts are in alignment.

I look closely at the trunnions, if the machine has any. Zinc die-cast parts (the very smooth ones with fine seam lines), like those used in some lightweight bandsaw and tablesaw trunnions, are not easily repaired. If these kinds of castings are broken, they should be replaced. If the machine is still in production, that's no problem. But if the manufacturer is out of business, you're stuck—unless you are willing to accept the expense of a custom-made part.

Make Sure the Motor Suits the Shop Wiring

Clean off the motor's data plate to see the kind of electrical power the machine requires. Three-phase motors rated at less than 5 hp usually can be swapped for ones that use standard single-phase current. The cost of replacing a motor is about $175★ for the first horsepower and $50★ for each additional horsepower. That doesn't include the price of a starter, cords, and plugs. Single-phase motors larger than 3 hp may require rewiring the shop to handle the heavy amperage load.

Some older machines use direct-drive, three-phase motors that can't be easily replaced (see the bottom right photo on the facing page). In that case, the shop must be outfitted with a phase converter. The cost can run from $100★ to $1,000★, depending on the size of the motor and the type of phase converter—static or rotary.

Will a Used Machine Need Work?

Count on it. The trick is knowing when potential repairs will cost more than the machine is worth.

MAKE SURE ALL THE PARTS are in the box before continuing your inspection. Finding missing parts for this out-of-production combination machine may be nearly impossible.

BADLY WORN PARTS may mean poor maintenance. Bandsaw blades are easy to replace, but a thrust bearing like this one indicates the machine was not carefully maintained.

REMOVE COVERS, and inspect power trains. The gears of this planer pass with flying colors–no cracks or broken teeth. If possible, try running the machine under a load to check the power train.

ESSENTIAL ADJUSTMENTS should work properly. Check the fence for squareness, run all the handwheels, and raise and lower the tables or blades. The fence shown here is out of square to the table.

CRACKED OR REPAIRED PARTS aren't necessarily a serious problem. Welding a frame is not difficult or costly. But care must be taken not to distort or misalign the parts.

SOME MOTORS CAN'T BE CHANGED. The three-phase, direct-drive motor on this jointer can't be swapped for a 110v single-phase motor. It will require 220v circuits and a phase converter.

I check the end of the motor shaft for mushrooming or peening damage from a hammer or gear puller that would make it difficult to install or remove a pulley. Then I grab the shaft to see if there is any up-and-down play. If so, the shaft, bearings, or bearing housing may be worn. If the motor is functional, I remove the drive belt and run it at idle. I listen for the hiss of the internal fan. If there's a high-pitched, gravelly whine when the motor coasts to a stop, the bearings need replacement.

The Power Train Must Run Freely

After checking the motor, I look at the power transmission components of the machine. I don't worry about the belts; I just assume they will need to be replaced. I look for pulley wobble, which means either the pulley bore or the shaft is worn. A new pulley isn't a major investment, but a worn shaft can be expensive to replace. Then I work my way through the power train, checking each part for wear.

Heavy-duty antique machines with babbitt bearings are often sold at bargain prices. Babbitt bearings work fine, but they can be difficult to replace and must be oiled before every use. The inevitable oil drips and spills can foul the surface of a board and may be a fire hazard. I tend to stay away from machines with babbitt bearings because newer machines with ball bearings are generally better made. However, with a little care, these machines are a good buy for an amateur millwright with the time and inclination to pour new bearings. Sometimes a machine with babbitt bearings can be modified for modern, standardized ball bearings, but it's no easy job.

I check all the ball bearings in the power train. I'm not concerned if some need replacement. Ball-bearing sizes are standardized, so it's easy to find replacements. However, pressing the bearings on and off the shafts requires some specialized tools. To do the job right, you'll need an arbor press and bearing pullers.

You can inspect gear-drive systems by simply removing the cover (see the top right photo on p. 131). One or two broken teeth can be repaired, but more than that will probably require a new gear. Some gear systems operate in an oil bath in an enclosed housing. In that case, I turn the machine over by hand to look for jerky motion in the visible moving parts and listen for grinding noises.

If possible, I run the machine under load and use it to cut some stock. Doing so can reveal problems, such as planers that don't feed, blades or belts that won't track, or excessive vibration due to damage or wear.

Moving the Machine Is Not Always Easy

It can take a significant amount of time and money to move a machine, and I keep that in mind while I'm inspecting it. A small machine like a Delta 14-in. bandsaw that weighs about 150 lbs. can be carried by two people to a van or pickup truck, but an old 36-in. Crescent® bandsaw is another matter. Sometimes I've moved large machines by disassembling them. I've also rented panel trucks with lift gates to carry heavy tools (but be warned that most lifts have a 1,000-lb. limit). I don't use lift gates when loading machines like drill presses, bandsaws, and mortising machines because these tall machines are dangerously unstable as the gate rises.

I've found the easiest way to move heavy machines is to hire a roll-back wrecker. It has unlimited headroom for tall machines and comes with an experienced operator/driver who can make moving a big machine much safer. The cost of moving a machine is about the same as moving a car over the same distance.

* Please note price estimates are from 1997.

ROBERT M. VAUGHAN is a contributing editor to *Fine Woodworking*. He repairs and restores woodworking machines in Roanoke, Virginia.

Restoring Vintage Machinery

BY ROBERT M. VAUGHAN

What a bargain—a 16-in. Walker-Turner bandsaw for $80. All it needed was new tires, guides, motor, electricals, guards, stand, complete disassembly, cleaning and rust removal, one casting weld, repainting, reassembly, and, of course, realignment of all the parts during reassembly. The good news was that all of the crucial components were there and in good condition; the other stuff I could fix. This wasn't an $80 bandsaw, but an $80 bandsaw kit. It was up to me to turn it back into a bandsaw.

I had to weigh the value of the restored bandsaw against commercially available machines. A resaw capacity of 12 in., 400 lbs.

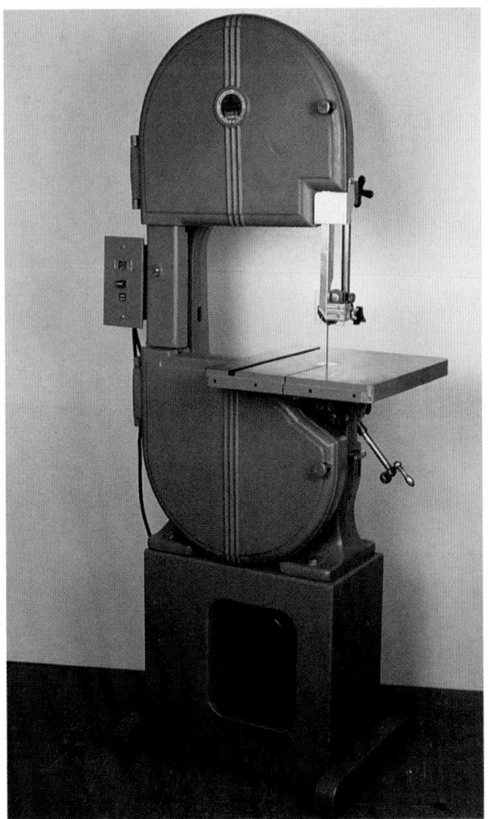

BEFORE THE RESTORATION, this old Walker-Turner wasn't much more than a pile of scrap iron. Sixty-five hours and four hundred dollars later, the author has a new (again), better-than-from-the-factory, vintage bandsaw. The saw's quality casting, 12-in. resaw capacity and 16-in. throat depth were all factors that made it worth restoring.

A FINE-WIRE BRUSH MOUNTED on the author's wood lathe quickly and efficiently cleans away dirt, dried grease, and even light rust. The wheel also imparts a slight polish, so Vaughan runs all fastener heads under the wheel for a few seconds.

MASKING ALL PARTS ENSURES a clean, crisp, professional-looking paint job. A good way of masking holes is to wrap a piece of paper tightly around a dowel and then to release it inside the hole. The paper will expand to fit. To avoid a messy cleanup later, remove all the masking tape as soon as the paint is dry enough to touch.

of quality American cast iron and a 16-in. throat depth are all factors that made this moderate-sized machine worth restoring. If this had been one of Walker-Turner's 14-in. models, I would have passed. The work required to restore it would have been the same, but the result would have been little better than a new Powermatic or Delta 14-in. model.

If you're thinking of restoring an old machine, it's important to realize that it's a very rare old machine that's ready to run. Almost all are like this machine was—a lot of cast iron with potential. Bearings, belts, pulleys, switches, wires and motor almost always need replacement. One reason that bandsaws are so popular to restore is that the parts that wear out can almost always be obtained from sources other than the original manufacturer. The important question to ask before diving into a restoration is whether the restored machine will be worth your trouble.

In this article, I'll discuss the general procedures common to restoring any old woodworking machine, as well as the more specific procedures that were necessary to get this bandsaw back into top form. And while the general procedures are applicable to just about any machine restoration, even the bandsaw-specific procedures illustrate ways of addressing problems common to all woodworking equipment—ways, for example, of dealing with dust, alignment problems, and beat-up or missing guards. The principles of machinery restoration are the same regardless of the machine.

Moving the Machine

Moving any heavy machine from one shop to another is always a chore. There are no rules on how to accomplish it other than to be prepared. I have help on hand for lifting. I generally bring resealable plastic bags for nuts and bolts and a note pad to record the disassembly sequence and to label parts bags. I also bring wrenches and WD-40 for disassembly of any heavy or protruding parts that might impede handling. I often remove the table, and any guards or pulleys, and I always try to remove the motor and cord. I make sure there are a couple of floor floats (four casters on a piece of plywood) ready in my shop, so I can move the machine around during the restoration process.

First Inspection

Once in the shop, I break out the air hose to blow out the years of accumulated dust and grease. Always wear a face mask or safety glasses when using an air nozzle. A 100-lb. blast of air into any of those little nooks and crannies can unleash hostile projectiles at bullet-like speed. If you don't have a compressor, a stiff bristle brush will remove most of the crud.

After I've cleaned off the bulk of the dust, dirt, and grease, I begin disassembly, examining each component for further mechanical problems—things I may have missed when I bought the machine. Organization at this stage really pays off. As I take apart the various subassemblies of a machine, I use plastic trays, bins, or boxes for the larger parts and resealable plastic bags to hold the little stuff. I note the sequence of washers, springs, and other things that I'd otherwise forget. I bag individually any shims I find, along with a note showing where they came from. This not only makes reassembly infinitely easier but also allows me to move the multitude of parts and store them out of the way, without losing track of what's what and what goes where—no little consideration in a space-starved shop.

Next I buy or collect all of the big items I'll need. This includes the motor, wiring, switch, pulleys, and belts—all the big-ticket items crucial to completion of the machine. Even when other unexpected expenses crop up, I know that the project will get finished.

FORTY YEARS OF DUST, dirt, and resin had taken their toll on the back blade guides, but they weren't damaged–just frozen. Vaughan removed the bearing from the shaft with two screwdrivers (left), popped the cap off the bearing with a hammer and dowel (bottom left), and sprayed the bearing clean with lacquer thinner (bottom right). The cleaned guide works like new.

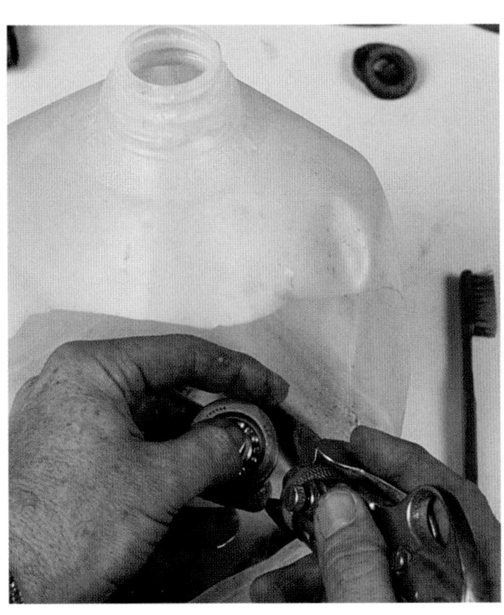

Cleaning

Proper parts cleaning is the most time-consuming aspect of the restoration process, but it's also the most important. The purpose of a thorough cleaning is not only to please the eye but to make things work as they should. I've been hired to repair a lot of equipment that needed nothing more mechanically challenging than a good cleaning. Forty years of dust, dirt, and resins have a way of adversely affecting the performance of the finest machine.

After covering my lathe bed (to protect it from flying dirt and debris), I mount a fine wire wheel on my wood lathe and use it to brush away any dirt or grit in threaded parts, to remove minor coatings of rust, and to clean up any dried, caked-on grease (see the photo at left on p. 134). The wire wheel also polishes a bit, so I put the heads of all the old screws, nuts, and bolts under the wheel.

I clean holes, with or without threads, with a brass brush (the kind used to clean rifle barrels) chucked into my electric drill. If the hole isn't very deep, I'll follow this with a blast of air and then wipe with a clean or solvent-dampened rag.

Grease is best removed with a solvent; I prefer lacquer thinner because it's the quickest solvent I normally have around the shop. I spray-clean small parts, using a compressor-powered spray gun and

How to Build a Good Machinery Stand

Constructing a well-made wooden stand is the single most important thing you can do to eliminate bandsaw vibration. A good wooden stand is superior even to a steel stand because the wood absorbs much of the vibration rather than transmitting and amplifying it as steel will. Another advantage of a wooden base is it's easily modified to accept hanging accessories, such as fences and miter gauges.

I build my stand first because I'd rather not be bending over for the whole restoration. Placement of the motor and electricals, provision for sawdust evacuation, ventilation, and machine maintenance are all factors I take into consideration when designing a stand.

The keys to a good stand are good materials and good construction methods. I use strong, dry hardwood and good-quality birch-veneer plywood. My construction methods are neither esoteric nor showy. Glued butt joints work fine as long as you use enough glue, and the joints fit tightly to begin with.

I push a joint together with the nose of my pneumatic staple gun, just until the glue begins to ooze out, and then I pull the trigger, squirting a 2-in.-long, wide-crown staple into the wood. The staple isn't for strength but rather to hold the joint tight until the glue dries. Using staples or nails without glue results in joints that are guaranteed to vibrate loose. Bolts are forever having to be tightened because of machine vibration and seasonal wood movement. Glued joints are the only sure way I know of to make vibration-resistant wood joints. Biscuit joinery also works.

GLUED JOINTS ARE THE KEY to building a good machinery stand, but lag bolts are good insurance against failed joints where the legs meet the carcase. Elaborate joinery isn't necessary for a strong and sturdy stand.

I glued the legs to the stand and then screwed large lag bolts through the leg, through the side of the plywood box, and into a backing block (see the photo on the facing page). The lag bolts compress the joint (pro-

spraying into a cut-out plastic milk jug. The milk jug catches most of the spray, which I use later to dampen rags for wiping down larger areas; I wipe with a dry rag after cleaning with a solvent-dampened rag. I've also found the refillable, rechargeable spray cans—which are available at most auto parts' stores—useful for cleaning larger areas. I just spray lacquer thinner on, then wipe clean with a dry rag. These cans are particularly handy in close quarters or when you don't want to drag the air hose around.

Think safety whenever working with solvents. Work in a well-ventilated area, wear a respirator, and always set the dirty rags outdoors—away from anything flammable—to dry after use.

Cleaning an old machine is messy work. Chances are that your workbench (and many other areas of your workshop) will become spotted with grease and grime. Make sure you clean up thoroughly after working on the machine before you begin working wood again. It's incredibly annoying to find greasy dirt smeared all over a just-completed project. Rebuilding a woodworking machine may not be as bad as rebuilding your car's transmission in your shop, but it's close.

Dirt or grease from a machine you're restoring can mess up your shop, but shop dust and dirt can mess up a restoration as well. To avoid this, make sure any surfaces you'll be working on are clean before you begin. Also try to finish the restoration without interruption. If you have to put your restoration on hold in midstream to work on a woodworking project, both can suffer unless you're extremely careful about cleanup and protection.

viding clamping pressure) and add a comforting margin of safety against joint failure. I also glued in a panel to accept tracks for sliding motor mounts and drilled and installed T-nuts in the top of the stand to accept the bolts that would connect machine and stand.

Since no floor is really flat, I put in T-nuts and steel leveling feet. Distributing the machine's weight equally to each of the four feet is essential to reducing vibration. Why don't I put locking casters on the base? The bandsaw is tall, thin, and has a high center of gravity, making it inherently subject to balance problems. Leaving a bandsaw permanently mounted on a wheeled base is trouble waiting to happen.

–R.V.

Renewing the Table

To clean up the rust on the tabletop, I started with 220-grit sandpaper wrapped around a block of wood, then moved up to 320 grit. After finishing with the 320 grit, I dampened the table with naphtha and rubbed with a hard Arkansas stone until the high spots shone like little mirrors. This makes any metal very slick and does wonders for planer and jointer beds—even new ones. It only has to be done once, and the results are well worth it.

General Machinery Repairs

Some repairs are specific to individual machines; others are general and apply to most machinery. I'll discuss general repairs below and the specifics of bandsaw repair in the sidebar on pp. 140–141.

All four wheel bearings in this saw were contaminated with dust and dried-out grease and needed to be replaced. The top bearings were standard sized and available locally (check the Yellow Pages for a bearing distributor near you), but the bottom bearings had an odd-sized inside dimension. My usual local sources of power-transmission products couldn't locate replacement bearings. I knew that Walker-Turner had some of its bearings custom-made for them, so I began to worry. I called Accurate Bearing Co. (see "Sources") and asked the sales manager about my bearings. He replied, "Sure. I have them right here. What else do you need?" I liked that.

To restore the outside threads of beat-up fasteners that can't easily be replaced, I used a thread-restoring file. These square files come in two sizes with eight different threads-per-inch sizes on each file. I set the file's teeth into the matching grooves of the fastener and filed. These files are particularly handy when the end of a threaded piece is smashed and when trying to start a threading die would risk cross-threading. You can find these files in most large industrial-supply catalogs.

The pulleys on the saw were cheap aluminum ones that no longer ran true. I replaced them with cast-iron pulleys from a local power-transmission distributor. The belt was equally worn, so I replaced it with a Browning® cogged, high-strength industrial belt (from the same distributor) that's designed to transmit high torque smoothly. Any machine is only as good as its weakest component, so these simple substitutions of power train components really make a big difference in the overall performance of the restored machine.

Any time something is held in place by a setscrew, there's a good chance that the point of the setscrew will cause a crater. The raised sides of these craters will cause all kinds of difficulties in disassembly, often requiring gear pullers, presses, punches, or a big soft-faced hammer. I usually file down the crater edges with a super-fine file or honing stone before removing the part from the machine. This avoids galling the inside of a hole or housing as the part is withdrawn.

The parts on this saw that need to be removed or adjusted to change the blade were fastened with nuts, bolts, and slotted-head screws. Every time I wanted to change blades, I'd have to hunt down the proper tools, have the tools and all loose hardware laying around during the blade change, and then put them all back when I finished. To make the machine more user-friendly, I replaced common nuts with wing nuts, bolts with threaded studs, and slotted screws with socket-head (Allen) screws. I then mounted a holder for the Allen wrench on the machine. Now I can change the blade and adjust the guides without ever going on a tool hunt.

Painting

Repainting a restored machine may deter rust, but the real reason is that it looks nice and makes you feel better about your machine. Sawdust may come off slightly easier, but who are you kidding?

How far you want to take the paint job is up to you. I've stripped down to bare metal, done body work, and built up the paint as though I was restoring an auto; other times, I've only needed to do touch-up work. Stripping may be necessary if the machine came from a school: often the color scheme will look as though it were designed by Stephen King and applied by King Kong. If you strip down old cast iron, you'll sometimes find that auto body filler was used to make a smooth surface.

On this particular machine, the existing paint had faded to olive-gray. I found original paint on an unexposed section of the machine and matched it with Krylon's® #1608 Smoke Gray. It took five cans to do this bandsaw, including the stand. I didn't bother to strip because the paint film was in good condition. I simply cleaned the

Sources

Carter Products
437 Spring St. N.E.
Grand Rapids, MI 49503
616-451-2928
www.carterproducts.com

Accurate Bearing Co.
1244 Capital Dr.
Unit 1
Addison, Ill. 60101
800- 323-6548
www.accuratebearing.com

Grainger®
800-473-3473
www.grainger.com
for branch locations.

surfaces with soap and water and then wiped them down with lacquer thinner. I had to spend a little more time and use a bit more solvent in some of the greasy corners and crevices, but there were no real trouble spots.

I mask all surfaces that take working parts, like shaft holes and ways. An easy way I have of masking the inside of a hole is to cut a small piece of paper and wrap it around a dowel. I then insert this into the hole and unwrap the dowel until the paper springs out to fill the hole.

I remove all masking tape and paper as soon as the part is dry enough to handle, so I won't have to deal with any sticky residue later. I paint the parts individually while they're disassembled. Bright, unpainted fasteners, new aluminum guards, and crisply contrasting parts, such as handwheels, all add up to create a quality impression. A wash-over paint job says something else altogether.

Electrical

This machine, like many older machines, had a simple toggle switch inconveniently located on the front of the frame. I replaced it with a new heavy-duty, push-button manual starter on the column (where it's easy to get to), but I had to cut a sheet-steel mounting plate for the switch first. After cutting and drilling the necessary holes in the mounting plate and mounting the starter, I drilled and tapped two holes in the bandsaw's cast-iron frame and attached the mounting plate assembly.

A rigidly mounted motor greatly reduces vibration. To mount the motor securely on this machine and still allow for tensioning of the belt, I cut a couple of short sections of folded steel U-channel (I used Unistrut® from Grainger; see "Sources"), drilled five holes in the bottom of each, and screwed them to the base. Then I found a couple of pieces of steel that would slide in the channels and drilled and tapped them to accept the motor.

A good-quality new motor, switch, wire, and plug will cost $200* to $300*. It's money well spent. I've used a light switch, vinyl-covered cord and a cheap mail-order motor before. Performance was poor from the start. I ended up shelling out more money for the good stuff in no time.

Bottom Line

Total material costs were just under $400*, bandsaw included. Costs were so low because I used a reconditioned motor and a manual starter (both good quality but without any bells or whistles) and because I already had just about all the peripheral materials (sheet steel and aluminum, clear plastic, fasteners) on hand.

I also spent about 65 hours on the restoration. At $25 an hour, labor costs would be about four times my materials' cost—not out of line for this kind of project. I've explained how I overhaul a machine and, for the most part, the reasons why. I hope this both inspires and instructs others to restore old machinery because the result, when done well, is most gratifying. The adage "they don't make them like they used to" is true, but there's a reason for it. The sad and brutal truth is that most buyers of new woodworking machinery don't demand quality so much as they do low-priced look-alikes. The downward spiral in the quality of woodworking machinery is the result. "They don't make discriminating buyers and users of woodworking machinery like they used to" is probably a more apt phrase. But who can criticize the guy who's perfectly satisfied with a five-dollar socket set?

* Please note price estimates are from 1993.

ROBERT M. VAUGHAN is a contributing editor to *Fine Woodworking*. He repairs and restores woodworking machines in Roanoke, Virginia.

The Particulars of the Bandsaw

Many of the steps in the restoration of this bandsaw are just as applicable to a vintage jointer or planer as they are to the bandsaw. A good stand, new electrical and drive systems, and clean, well-lubricated bearings are things that any old machine can profit by. But the procedures below are bandsaw-specific, and although it's a Walker-Turner I happened to be working on, the steps taken—and the conditions that necessitated these steps—are common to most bandsaws.

TABLE, BLADE-GUIDE POST, BLADE

For a bandsaw to work properly, the blade, blade-guide post, and table must all be in proper relation to one another. If they're not, every time you move the upper guide up or down, you'll have to readjust the guides. You can either live with this long-term hassle or go through the one-time tedium of correcting these misalignments. This alignment has been adjustable on every bandsaw I've ever worked on, but I've never worked on inexpensive do-it-yourself-type bandsaws, so regretfully, I have no experience in that realm.

On C-frame bandsaws, such as this Walker-Turner, the upper blade-guide post's line of travel is dictated by the position of a hole in the casting. To make the blade travel in a line parallel to that of the post, the position of the wheels needs to be adjusted properly. On this machine, the upper wheel can be moved from side to side. I was also able to make slight adjustments to the position of the whole upper wheel carriage by loosening and retightening its mounting bolts. The bottom wheel can be moved in and out by adjusting the setscrew-held bearing stops. It took quite a few adjust-tighten-test sequences before the blade's line of travel was right (parallel to the upper blade-guide post), but now I won't ever have to worry about it again.

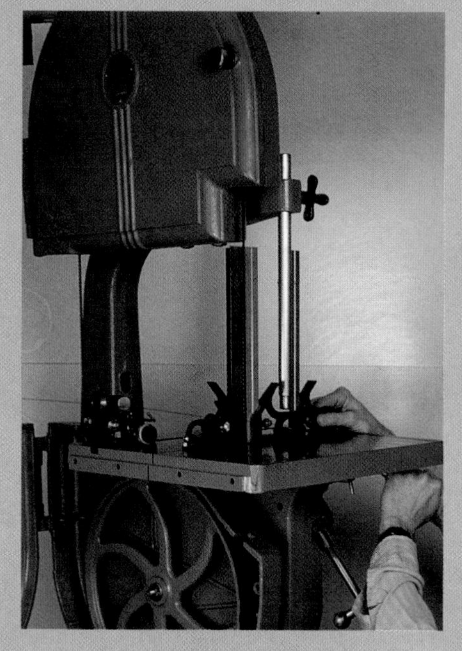

THE SAW TABLE MUST BE PERPENDICULAR to both blade and upper blade-guide post (which must be parallel to each other) for the bandsaw to work perfectly. The author adjusts the table for perpendicular, using combination squares to check for gap between the rule and the blade and between the rule and the blade-guide post.

The next step was to set the table perpendicular to the blade's line of travel. I first put some good squares against the sides of the blade, loosened the tilt lock, and tilted the table until both blades were parallel with the sides of the blade. Next, I put the squares' blades against the front and back of the blade and loosened the underneath bolts that hold the table to the trunnions. I slipped thin sheet-metal shims between the table and the trunnions, experimenting with different thickness shims until the table was perpendicular to the blade. If you can't get the table perpendicular, then all you can do is split the difference on either side of the blade. Once again, though, it's a one-time hassle with long-term benefits.

RESTORING THE OLD BLADE GUIDES

Deep grooves were worn into the side guide blocks, and the back guide bearings were virtually frozen from sawdust contamination. Using a disc sander and miter gauge, I reground the 45° angle on the steel side blade guides.

MITER GAUGE, DISC SANDER, and a mitered push stick allowed Vaughan to grind the side blade guides with little effort. Other options, had the guides been irreparable, would have been to grind his own from steel bar stock, to replace them with new Delta guides or with guides made of graphite-impregnated phenolic resin (sold as Cool Blocks by Garrett Wade).

For both top and bottom back guides, I clamped the guide bearing's shaft in a vise and used two large opposing screwdrivers to pry off the bearing. I reversed the bearing and drove off the front dust cover with a wooden dowel, exposing the balls and cage. Then I flushed the bearings clean with lacquer thinner. Don't let the bearing spin freely under a blast of air because solvent will spew everywhere, and the bearing can be damaged at high speeds.

If either the side or back guides had been irreparable, I would have had to replace the defective part. For replacement side guides, I could have used graphite-impregnated phenolic resin guides (Cool Blocks), cut and ground new guides, or searched for compatible guides from another manufacturer. If the back ball-bearing guide for this particular machine had needed to be replaced, I could have had a new shaft machined to fit a standard 6203 size bearing—no big deal for someone with a metal lathe.

A NEW BLADE GUARD

Blade guards on old bandsaws, if they're there at all, are rarely in good working order, and this machine was no exception. After studying the old blade guard, with its worn-out sliding pieces, I decided that I could make one that would work better and be easier to maintain. The only hitch was that I had to have an aluminum block machined to accept the bent sheet-aluminum guard. Knowing a tall, single piece guard fastened to the upper blade guide could not fully extend up into the castings enclosing the upper wheel, I had to make two separate guards. One short guard was fastened to the upper wheel guard casting. It came down about even with the bottom of the upper wheel guard castings and was wide enough to let the whole upper blade guide slide up behind it. A tall, skinny guard was fastened to the upper blade-guide area. At the blade guide's lowest position, the top of this guard did not come below the guard I put on the upper wheel casting. This way, I could raise the blade guide all the way up and prevent the tall guard on the blade guide from hitting things in the upper wheel guard castings. As an afterthought, I put a clear plastic panel on the front of the blade-guide mounted guard to provide extra protection on the feed side of the saw. I made this panel even with the actual bottom of the blade guard so that any stock that would fit under the plastic panel would fit under the upper blade guide.

OFTEN THE SIMPLEST SOLUTIONS are the best. The author had noticed that the bottom back blade guides are almost always inundated with dust and tend to lock up much more quickly than the top ones. By placing an angled piece of clear plastic between the saw table and guide, he was able to deflect almost all of the dust away from the guide.

A NEW GUARD OF BENT SHEET-ALUMINUM seemed an easier and better solution than trying to get the rusted old steel guard (left) back into shape. Vaughan added the clear plastic to the front of the guard for more protection on the saw's infeed side.

CUSTOM DUST DEFLECTORS

Bandsaws normally flood the bottom blade-guide assembly with sawdust, causing these guide's bearings to fail long before the top guide bearings. Also, since the bottom guides are hidden from convenient inspection, they're rarely cleaned. I was able to eliminate more than half of the normal sawdust deluge with a clear plastic deflector, which I mounted above the guide. I experimented with cardboard and tape until I had a good pattern, and then I cut out what I needed from some scrap clear plastic. I bent, drilled, and tapped a strip of aluminum to hold the plastic deflector. I then drilled and tapped one hole in the bottom wheel guard and threaded a bolt through from the inside. I fastened the deflector assembly with a wing nut for easy removal and for when I needed to change blades or set the saw table at a particularly steep angle.

Even with the plastic deflector, I noticed dust being broadcast from the joints of the clamshell castings of the lower wheel guard. This dust would blow out on the table and pile up on the floor behind the machine. I found I could direct most of this dust toward the front of the saw with two simple baffles. After experimenting with cardboard again to get the best pattern, I cut baffles from aluminum flashing and then mounted them to the guard by drilling and tapping a single hole for each baffle.

The net result of all of this activity is that most of the sawdust created by the saw winds up right in front of it, where it can be easily swept away. If I used a vacuum system, I'd have cut and mounted a piece of 2-in. tubing in place of the plastic deflector to suck up the dust right as it comes off of the blade. —R. V.

Protecting Surfaces in the Shop

BY CHRIS A. MINICK

We go to great lengths to protect the projects that leave our workshops from the rigors of everyday use, yet we often neglect our jigs, fixtures, shop cabinets, workbenches, and tools—the very objects that allow us to create fine furniture in the first place. There's no need to French-polish your crosscut sled, but there are treatments that will make your jigs, benches, and machine tops work better and last longer.

Prevent Rust on Machine Surfaces

Bare metal will rust if not protected or used constantly. There are myriad products that purport to be the last word on metal

Wax for Machine Surfaces

A green pot-scrubbing pad mounted to the hook-and-loop base of a random-orbit sander (left) removes rust and tarnish from machine beds. Kitchen waxed paper (below) can be used to protect and lubricate a machine bed. Simply crumple up the paper and rub it over the surface (right) to lay on a thin layer of wax.

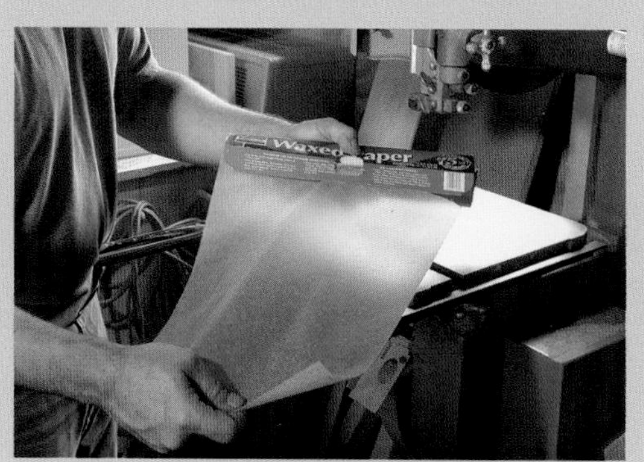

protection. Each product claims to prevent rust and to leave a slick surface that will not contaminate wood. The ones I've tried have done neither, and I'm leery of applying oil to my saw top or jointer bed. As a consequence, I've adopted my own system for cleaning and protecting metal that works and doesn't require much effort.

I remove the rust by buffing the surface with a green pot-scrubbing pad. Small items such as planes, chisels, or scrapers are buffed by hand, but large surfaces like my tablesaw or jointer get power-buffed with a 5-in. random-orbit sander. The hook-and-loop sanding pad holds the scrubber firmly while I guide it around the metal surface. Once it's rust-free, I polish the surface with a wadded-up piece of waxed paper. Just enough wax (carnauba, I've been told) transfers to the metal surface to protect it from rusting and to lubricate it simultaneously. Boards glide across the saw or the jointer as if they are floating on air.

High-pressure laminate requires virtually no maintenance and is the ideal surface for router tables, shop-cabinet tops, chopsaw extensions, and the like. Dried glue and finish easily scrape off, and a quick rubdown with waxed paper keeps the laminate slick.

Finishes Help Jigs Last Longer

It may seem unnecessary to finish jigs, but there are several good reasons to put in the effort. Even simple jigs help us hold, saw, or bore wood precisely. The usefulness of most jigs depends on their accuracy, and they require a fair amount of time and effort to construct, so it makes sense to protect them properly. The type of finish depends on how the jig is used and the material it is made from.

Shellac on Solid Wood Finishing jigs with shellac virtually eliminates dimensional changes in solid wood caused by seasonal humidity. Shellac has the lowest moisture-

Finishes for Jigs

SHELLAC

Shellac not only seals and lubricates, but it's also the best finish for reducing seasonal movement in solid wood.

WAX

The plywood parts of jigs need only a coat of paste wax to protect them from dirt and glue drips. The wax also helps jigs slide during use.

GLUE SIZE

The edges of MDF are far more absorbent than the compressed faces. PVA glue diluted with water (called glue size) seals the edges, after which the whole jig can receive a coat of shellac.

Two Ways to Protect Benchtops

SACRIFICIAL SURFACE
Benchtops made from sheet goods are best protected with a sacrificial top of hardboard, which can be replaced easily when it gets worn out (left). Use the old surface as a template to mark the outline on the new top (below).

vapor transmission rate of any finish on the market today. That means the moisture content of wood finished with a few coats of shellac will barely change, even if the level of humidity in the shop varies significantly.

I prefer dewaxed shellac to finish furniture, but for jigs I use the hardware-store variety of premixed orange shellac. That cloudy stuff in the can is natural shellac wax, which lubricates the sliding surfaces of the jig and allows it to glide across a tablesaw easily.

Paste Wax for Plywood Plywood and tempered hardboard remain dimensionally stable over wide humidity swings, so they are ideal materials for your most critical jigs. Often a coat of furniture paste wax is all the protection needed for jigs made

A RENEWABLE FINISH THAT SEALS AND PROTECTS

SCRAPE FIRST. On a used bench, scrape away any dried glue before applying a new coat of the wipe-on finish.

MIX THE FINISH. Minick's mixture of oil, varnish, and paint thinner contains a higher proportion of resin than most Danish oils to give the bench greater protection.

LAY IT ON HEAVY. Apply the oil-varnish mixture liberally and allow it to soak in for 10 minutes before wiping off the surplus with a clean cloth.

from these stable materials. The paste wax also helps the jig slide more easily during use.

Glue Size Seals MDF Medium-density fiberboard (MDF) is less seasonally stable than other manmade sheet goods, but it still can be used for jig construction if properly sealed. The edges of an MDF sheet are considerably more absorbent than the faces. If left unfinished, the unprotected edges can absorb moisture and significantly swell the MDF sheet. A coat or two of finish will not seal the edges adequately, so instead I use glue size—a 50-50 mixture of polyvinyl-acetate (PVA) glue and water. Allow the glue size to dry completely, lightly sand the rough surface, then finish the whole jig with shellac or lacquer. This will provide an effective moisture barrier.

Hold the Varnish For finishing jigs, I prefer shellac or cabinet-grade nitrocellulose lacquer, which dry much faster than oil-based varnish and are available in convenient aerosol cans. More important, shellac and lacquer form harder surface films than varnish, which means that jigs coated with these finishes slide more easily over a tablesaw or router table than their varnish-coated cousins.

Benchtops Need Protection, Too

When it comes to finishing a workbench, I find that woodworkers fall into two camps: Some lavish as much attention on finishing their bench as they would a dining-room table, while others consider any finish to be a waste of time. I fall somewhere in the middle, applying a finish more for protection than for the look it imparts.

Benchtops take plenty of abuse, so an easily renewable finish is in order. My home-brewed wipe-on finish dries fast and gives plenty of protection. Mix 1 cup oil-based polyurethane brushing varnish with ½ cup mineral spirits, then add about 2 oz. boiled linseed oil. Adjust the mixture with mineral spirits for proper wiping consistency, and then apply it like any other wipe-on finish. It will dry to the touch in about one hour. Three coats on a new workbench will resist water better than standard Danish oil; it's a good idea to apply an additional coat or two once a year to maintain this protection.

My shop cabinets are varnished inside and out, so they don't need much attention; however, they still get dings and scratches. Every couple of years, I vacuum off the dust and wipe on my home brew, which makes them look almost as good as new.

If your benchtop is made from layers of plywood or MDF, a better form of protection is to use a sacrificial surface. My benchtop is covered with ⅛-in.-thick hardboard held in place with double-faced carpet tape. When the top gets too dinged to be usable, I pry up the old piece and replace it with fresh hardboard.

CHRIS A. MINICK is a consulting editor to *Fine Woodworking* magazine.

Taming Woodworking Noise

BY JACK VERNON, PH.D.

Those tools that we most need to use are the very ones that offer the greatest potential danger to our ears. Common woodworking machines such as routers, planers, and tablesaws can cause permanent hearing damage. The good news is that there are easy ways to protect your hearing while continuing to work wood with power tools. But first it helps to understand the problem.

Hearing-Damage Basics

Loud sounds damage hearing in much the same way that earthquakes damage buildings. Loud sounds simply shake apart the delicate inner-ear structures called hair cells. These hair cells are highly specialized nerve endings designed to receive sound energy and convert it into neural impulses. In turn, those neural impulses produce our ability to hear. Sound waves strike all parts of our body, but only the hair cells of the inner ear can convert that sound energy into what causes us to hear. Once destroyed, the hair cells are gone forever.

Open a pea pod, and take out one pea. That is about the size of your inner ear, and amazingly, it contains 30,000 hair cells and approximately the same number of nerve fibers leading away from the hair cells up to the brain. You can easily appreciate that the inner ear is not only a very delicate structure, but it is compacted into a very small space.

Measuring Woodworking Noise

As might be imagined, damage to the ears produced by loud sounds is a combination of the intensity of the sound and the length of time to which one is exposed to that sound. Sound measurements of what we commonly call loudness, which actually measure the sound pressure level (SPL), are expressed in decibels (dB). Decibel units represent a logarithmic scale because the human ear can perceive such a large range of different inten-

PROPER EAR PROTECTION HELPS PREVENT HEARING DAMAGE. Common woodworking machines, like this table-mounted router, can quickly damage hearing unless you protect your ears. LeRoy Schmidt, foreman of the carpenter shop at Oregon Health Sciences University, wears ear muffs during tests conducted by the Oregon Hearing Research Center.

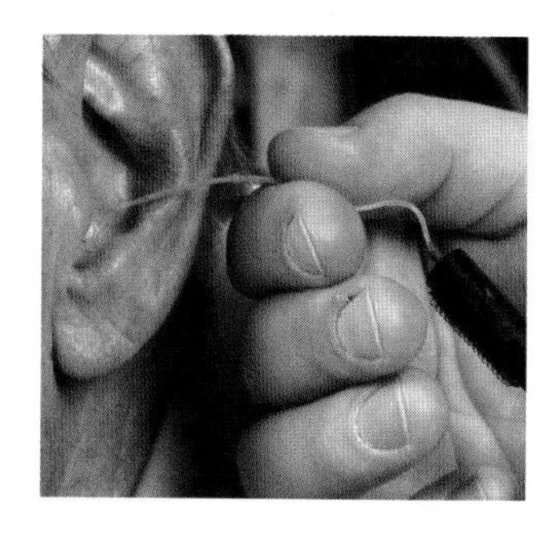

THE EAR IS AN AMPLIFIER. Using this miniature microphone inserted into the ear canal,researchers measured sound levels right at the ear drum. The findings showed the ear canal amplifies woodworking noise on average by about 7 dB over conventional readings taken at ear level.

sities. Logarithmic notations are expressions of ratios; for example, if one sound is twice as intense as another sound, it is 6 dB more intense. If one sound is 10 times more intense than a second sound, the first sound is 20 dB more intense.

The federal Occupational Safety and Health Administration (OSHA) standards limit industrial workers to 90 dB SPL for an eight-hour day. For sound levels of 95 dB, only a four-hour work day is allowable. At a sound level of 100 dB, the allowable work day is two hours, for 105 dB one hour, 110 dB 30 minutes, 115 dB 15 minutes and so on.

To study how woodworking machines affect hearing, members of the Oregon Hearing Research Center staff measured the intensity (loudness) of the sound in the conventional manner, at the ear level. But we also used special equipment—a miniature microphone—to measure sound intensity inside the ear canal at the ear drum itself (see the photo at left). The chart lists our measurements of typical woodworking machines under appropriate and normal operating conditions. Keep in mind, however, that tools vary from maker to maker with some being louder than others.

The sound measured at the ear drum is significantly louder than that same sound measured at ear level. In practical terms, that means that the ear canal leading to the ear drum produces some amplification of the sound intensity. For example, the noise generated by the 15-in. planer was 9 dB more intense at the ear drum than when measured at the ear level. On average, the sound produced by the machines we measured was increased by the ear canal about 7 dB (and remember an increase of 6 dB is a doubling of the sound intensity). In other words, sounds are more than twice as loud at the ear drum.

Another problem with woodworking power tools is that when we hold them, we stimulate our ears by bone conduction (sound traveling through the body), as well as by airborne sounds. Ear muffs and ear plugs block out sound coming to the ear through the air. When bone conduction of sound is involved, it would be desirable to use the power tool in short bursts to minimize any accumulation effect. Antivibration gloves may also help, but we have not tested them.

HOW LOUD ARE WOODWORKING MACHINES?

Type of Machine	Sound Intensity	
	At ear level	At ear drum
Nail gun (6d 2-in. finish nail)	104 dB	110 dB
Chopsaw	102 dB	108 dB
Router	104 dB	107 dB
15-in. planer	96 dB	103 dB
10-in. tablesaw	95 dB	103 dB
Palm sander (quarter sheet)	96 dB	103 dB
Panel cutter	95 dB	102 dB
Dust collector	93 dB	99 dB
Bandsaw	92 dB	98 dB
Shop vacuum	90 dB	97 dB
10-in. tablesaw (with Silencer blade)	86dB	93 dB
6-in. jointer	80 dB	90 dB

Warning Signs of Hearing Damage

When other parts of our bodies are damaged, the warning signal is pain. But for the ear, the warning signal is tinnitus (ringing in the ears). Ringing in the ears after exposure to a woodworking tool means that tool was too loud for your ears and that you should always wear ear protection in the future when using that tool. Don't be guided by the actions of others. Some people have tough ears, and some people have tender ears, with all grades in between. You may have noted that Norm Abram of The New Yankee Workshop seldom wears ear protection. I would assume that Mr. Abram has tough ears.

The way in which hearing impairment starts can be deceptive, so deceptive as to go unnoticed initially. Imagine the inner ear laid out like a piano keyboard, the low frequencies to the left and the high frequencies to the right, with each frequency systematically spaced in between. It is the high frequencies that are damaged initially by loud sounds, so one can sustain a considerable amount of damage before the ability to hear the lower pitches becomes impaired. The sounds to which we pay attention and which we commonly use are restricted to the low-frequency portion of the ear, starting with about 4,000 Hz (cycles per second) and moving to lower pitches.

The typical course of hearing loss is something like this: With the initial hearing loss, the person has no difficulty hearing and understanding speech as long as the person is in a relatively quiet place. But when there is background noise present (in a restaurant or at a cocktail party, for example), the person will hear speech, but he or she will not be able to understand it. This condition is an early warning of hearing loss. Moreover, that kind of hearing loss often can be compensated by a pair of properly fitted hearing aids.

Ear Protection

There are two common forms of ear protection: ear muffs and ear plugs. In extreme cases, it is advisable to use both types at the same time. Much has been made of ear plugs as a protective device, and it is true that ear plugs work about as well as ear muffs. Ear plugs, such as the foam EAR brand plugs available in most drug stores, are good protective devices, and they are inexpensive. A disadvantage of ear plugs is that they can be difficult to insert correctly. And it takes time to get them inserted. In addition, the ear plug is vulnerable to jaw movements, which can break the sound seal. Place your finger in your ear and move the jaw, as in chewing or talking, and note the amount of ear canal movement. It is this movement that can make ear plugs less effective.

Properly selected ear muffs offer as much sound protection as do custom-fitted ear plugs. More importantly, ear muffs are easier to put on and take off, provided they are available at each noisy machine. If the muffs are on the other side of the shop from the tool being used, there's an inclination to say, "This is a very brief task; I don't really need ear protection." We recommend that a pair of ear muffs be placed on each machine capable of producing ear-damaging loud

HARMFUL SOUND ALSO CAN BE TRANSMITTED THROUGH YOUR HANDS. **Proper ear protection may not be enough when using hand-held power tools like this router. Bone conduction can transmit harmful noise through your body to your ears. Limiting the duration of use of such tools may help.**

TAKE OFF THOSE NOISY SAWBLADES AND REPLACE THEM. **Several manufacturers, including Freud, CMT, and Everlast (not shown) now offer special sawblades designed to cut down on noise. Changing to one of these blades may reduce saw noise by up to 7 dB. The blades have laser cuts that dampen noise and vibration.**

RUBBER MOUNTING PADS CAN CUT MACHINE NOISE.. **Mounting machines on antivibration mats or on rubber foot pads, such as those on this 1-in. belt sander, can cut down on woodworking noise by isolating or dissipating the vibration that causes noise.**

Sources

Safety and Supply
595 N. Columbia Blvd.
Portland, OR 97217
503-283-9500
Thunder 29 ear muffs

Everlast Saw and Carbide Tools, inc.
9 Otis St.
West Babylon, NY 11704
516-491-1900
Silencer sawblade

Oregon Hearing Research Center
3515 SW Veterans Hospital Rd.
Portland, OR 97201-2997
503- 494-8032

sounds. The ear bows of safety glasses can break the sound seal for some ear muffs, but our research showed the cuffs on the Thunder 29 ear muffs (see "Sources") are sufficiently pliable that they can be worn over glasses without any loss of sound protection.

Quieting Woodworking Machines

In addition to ear protection, it is possible to reduce the amount of sound generated by certain machines. Several manufacturers are marketing so-called quiet sawblades (see the bottom photo on p. 149). We tried the Silencer 10-in. sawblade (see "Sources"), and we found it reduced tablesaw noise level by 9 dB when compared to a regular carbide blade. Remember that reducing sound intensity by 6 dB means cutting the intensity in half; a reduction of 10 dB means a reduction of three times. Thus the 9-dB sound reduction provided by the Silencer blade is significant.

Many machines, such as the bandsaw, produce some of their noise by the resonance of their metal panels. Attaching pieces of plywood to these panels helps reduce the sound generated by the saw. For example, the noise produced by a 16-in. Grizzly® bandsaw was reduced from 92 dB to 89 dB by loading its panels with plywood. Mounting tools on rubber isolation blocks or wood mounts can reduce sound levels (see the photo above).

Keep tools in good working order. Dull tools tend to make more noise than do sharp ones. Misaligned belts and pulleys can generate excess drive-train noise. Worn or poorly lubricated bearings will add to noise.

Think about your hearing when you purchase woodworking equipment. Some designs and individual tools are louder than others. Machines with universal motors tend to be louder than those with induction motors. Gear-driven tools are usually louder than belt-driven or direct-drive tools.

In general, shielding, insulating, and muffling can reduce machine noise. The degree to which these procedures are effective depends in part upon the conditions of the individual shop, such as size, shape, surface of the walls, ceiling construction, and ceiling height. Each situation requires individual attention. But the point is to look for ways to reduce sound levels, and you will find them. John Culp of Peachtree City, Georgia, created a muffler for his two-stage dust collector using open-cell foam. He said the effect on the machine's performance was "negligible, but the high pitch whine is greatly reduced, making it much more comfortable for unprotected ears."

It's important to prevent permanent damage by protecting your ears from harmful noise any way you can. Remember: If after exposure to a noise your ears ring, even briefly, then the sound was too intense for your ears, and in the future, use ear protection.

JACK VERNON is director of the Oregon Hearing Research Center at Oregon Health Sciences University in Portland, Oregon. Jim Nunley and Jonathan Lay, also members of the research center, contributed to this article. All three men are active amateur woodworkers. Those who already may be suffering from tinnitus can contact the Oregon Hearing Research Center (see sources) to learn about relief procedures for tinnitus.

Credits

The articles in this book appeared in the following issues of *Fine Woodworking*:

p. 4: Flat, Straight, and Square by Peter Korn, issue 102. Photos by Vincent Laurence, © The Taunton Press, Inc.

p. 10: From Rough to Finish by Gary Rogowski, issue 131. Photos by Anatole Burkin, © The Taunton Press, Inc.; Drawings by Chris Clapp, © The Taunton Press, Inc.

p. 17: From Rough to Ready by Roger A. Skipper, issue 145. Photos by Asa Christiana, © The Taunton Press, Inc.; Drawings by Bob La Pointe, © The Taunton Press, Inc.

p. 25: Choosing a Tablesaw by Robert M. Vaughan, issue 112. Photos by Charley Robinson, © The Taunton Press, Inc., except photo on p. 25 © Robert M. Vaughan.

p. 32: Tablesaw Tune-Up by Kelly Mehler, issue 114. Photos by Jonathan Binzen, © The Taunton Press, Inc.; Drawings by Bob La Pointe, © The Taunton Press, Inc.

p. 39: Shopmade Outfeed Table by Frank A. Vucolo, issue 108. Photos © Alec Waters; Drawings by David Dann, © The Taunton Press, Inc.

p. 43: Jointer Savvy by Bernie Maas, issue 102. Photos © Alec Waters; Drawings by Mark Sant' Angelo, © The Taunton Press, Inc.

p. 50: Jointer Tune-Up by John White, issue 142. Photos by Anatole Burkin, © The Taunton Press, Inc.; Drawings by Vince Babak, © The Taunton Press, Inc., except drawing on p. 55 by Michael Pekovich, © The Taunton Press, Inc.

p. 58: Getting Peak Planer Performance by Robert M. Vaughan, issue 107. Photos © Robert M. Vaughan, except photo on p. 58 © Alec Waters; Drawings © Matthew Wells.

p. 66: The Jointer and Planer Are a Team by Gary Rogowski, issue 160. Drawings by Jim Richey, © The Taunton Press, Inc.

p. 71: All About Bandsaw Blades by Lonnie Bird, issue 140. Photos © Lonnie Bird, except photo on p. 71 by Michael Pekovich, © The Taunton Press, Inc.; Drawings by Vince Babak, © The Taunton Press, Inc.

p. 80: Shopmade Tension Gauge by John White, issue 147. Photos by Tom Begnal, © The Taunton Press, Inc., except photos on p. 83 (top and bottom left) by Michael Pekovich, © The Taunton Press, Inc.; Drawings by Vince Babak, © The Taunton Press, Inc.

p. 85: Bandsaw Tune-Up by John White, issue 157. Photos by Erika Marks, © The Taunton Press, Inc.; Drawings by Jim Richey, © The Taunton Press, Inc.

p. 92: Soup Up Your 14-in. Bandsaw by John White, issue 159. Photos by Tom Begnal, © The Taunton Press, Inc., except photo on p. 92 by Michael Pekovich, © The Taunton Press, Inc.; Drawings by Melanie Powell, © The Taunton Press, Inc.

p. 102: Jobs a Shaper Does Best by Lon Schleining, issue 112. Photos © Alec Waters; Drawings by Kathleen Rushton, © The Taunton Press, Inc.

p. 110: Choosing Shaper Cutters by Lon Schleining, issue 118. Photos © Robert Marsala, except photo on p. 110 © Alec Waters.

p. 115: Boring Big Holes by Robert M. Vaughan, issue 129. Photos by Anatole Burkin, © The Taunton Press, Inc., except photo on p. 115 by Michael Pekovich, © The Taunton Press, Inc.

p. 121: Jigs for the Drill Press by Gary Rogowski, issue 140. Photos by Anatole Burkin, © The Taunton Press, Inc.; Drawings by Vince Babak, © The Taunton Press, Inc.

p. 127: Buying Used Machinery by Robert M. Vaughan, issue 122. Photos © Robert M. Vaughan, except photos on p. 127 and p. 128 © Aimé Fraser, and top left photo and bottom center photo on p. 131 © Dennis Preston.

p. 133: Restoring Vintage Machinery by Robert M. Vaughan, issue 118. Photos © Robert M. Vaughan.

p. 142: Protecting Surfaces in the Shop by Chris A. Minick, issue 167. Photos by Mark Schofield, © The Taunton Press, Inc., except photos on p. 144 by Michael Pekovich, © The Taunton Press, Inc.

p. 147: Taming Woodworking Noise by Jack Vernon, Ph.D., issue 110. Photos © William Sampson.

Index

The New Best of Fine Woodworking series

A collection of the best articles from the last 10 years of Fine Woodworking.

OTHER BOOKS IN THE SERIES

Designing Furniture
The New Best of Fine Woodworking
From the editors of FWW
ISBN 1-56158-684-6
Product #070767
$17.95 U.S.
$25.95 Canada

Small Woodworking Shops
The New Best of Fine Woodworking
From the editors of FWW
ISBN 1-56158-686-2
Product #070768
$17.95 U.S.
$25.95 Canada

Working with Routers
The New Best of Fine Woodworking
From the editors of FWW
ISBN 1-56158-685-4
Product #070769
$17.95 U.S.
$25.95 Canada

Building Small Projects
The New Best of Fine Woodworking
From the editors of FWW
ISBN 1-56158-730-3
Product #070791
$17.95 U.S.
$25.95 Canada

Designing and Building Cabinets
The New Best of Fine Woodworking
From the editors of FWW
ISBN 1-56158-732-X
Product #070792
$17.95 U.S.
$25.95 Canada

Traditional Finishing Techniques
The New Best of Fine Woodworking
From the editors of FWW
ISBN 1-56158-733-8
Product #070793
$17.95 U.S.
$25.95 Canada

Working with Tablesaws
The New Best of Fine Woodworking
From the editors of FWW
ISBN 1-56158-749-4
Product #070811
$17.95 U.S.
$25.95 Canada

Working with Handplanes
The New Best of Fine Woodworking
From the editors of FWW
ISBN 1-56158-748-6
Product #070810
$17.95 U.S.
$25.95 Canada

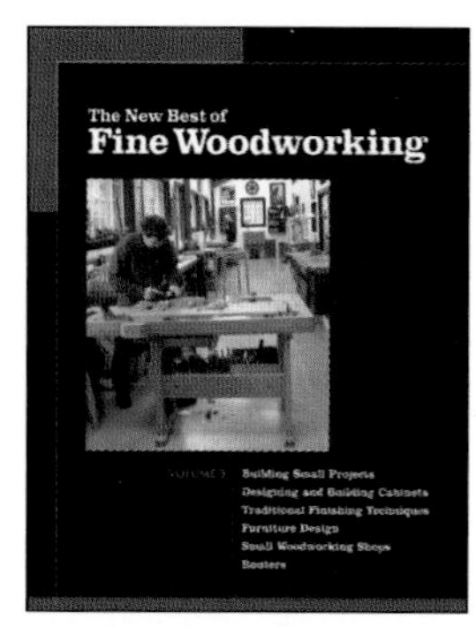

The New Best of Fine Woodworking Slipcase Set Volume 1

Designing Furniture
Working with Routers
Small Woodworking Shops
Designing and Building Cabinets
Building Small Projects
Traditional Finishing Techniques

From the editors of FWW
ISBN 1-56158-736-2
Product #070808
$85.00 U.S.
$120.00 Canada